Harris on the Pig
The Breeding, Rearing, Management and Improvement of Swine

by Jos. Harris of Moreton Farm

with an introduction by Jackson Chambers

Originally published by Orange Judd

This work contains material that was originally published in 1870.

This publication is within the Public Domain.

This edition is reprinted for educational purposes
and in accordance with all applicable Federal Laws.

Introduction Copyright 2018 by Jackson Chambers

Self Reliance Books

Get more historic titles on animal and stock breeding, gardening and old fashioned skills by visiting us at:

http://selfreliancebooks.blogspot.com/

Introduction

I am pleased to present another title in the "Raising Pigs" series..

As with all reprinted books of this age that are intended to perfectly reproduce the original edition, considerable pains and effort had to be undertaken to correct fading and sometimes outright damage to existing proofs of this title. At times, this task is quite monumental, requiring an almost total "rebuilding" of some pages from digital proofs of multiple copies. Despite this, imperfections still sometimes exist in the final proof and may detract from the visual appearance of the text.

I hope you enjoy reading this book as much as I enjoyed re-publishing and making it available to fanciers again.

With Regards,

Jackson Chambers

PREFACE.

Paradoxical as it may seem, in writing a book on Pigs and in endeavoring to show that we can obtain more meat from a well-bred pig, in proportion to the food consumed, than from any other domestic animal, it is no part of my object to stimulate the production of pork.

For over twenty years I have had the honor to be connected with the Agricultural Press of America, and have had my thoughts constantly directed to the means necessary to improve our general system of farming. A farmer's son, and myself a farmer, all my sympathies are with the farming class rather than with the consumers; but I am satisfied that, in many respects, our interests are identical. It should be our study to furnish good food at reasonable rates. At the present time the consumers in our large cities are obliged to pay much more for flesh-meat than it is intrinsically worth; and, on the other hand, with the exception of those who produce beef and mutton of the best quality, farmers make nothing by raising and feeding cattle and sheep. We receive more for our meat than it is worth, and yet it costs us more than we get for it.

The remedy for this unsatisfactory condition of affairs, will be found in cultivating our land more thoroughly, in growing better grass, in keeping better stock and in liberal feeding.

The introduction of better breeds of pigs will in itself do little towards improving our farms; but the farmer who once uses a thorough-bred boar and adopts a liberal system of feeding, will find that he can produce better pork at a far less cost than when he uses a common boar;

and he will be likely to study the principles of breeding with an interest he has never felt before. The introduction of a thorough-bred boar will lead to the introduction of a thorough-bred ram and a thorough-bred bull of a good breed, and this, in conjunction with cleaner culture and a more liberal feeding, is all that is needed to give us better and cheaper meat; and at the same time we shall make more and richer manure, and be enabled to grow larger and far more profitable crops of grain.

I believe I was the first writer who contended that, other things being equal, it was desirable to get animals that would eat, digest and assimilate a large amount of food. In the following pages I have endeavored to give some reasons for this opinion and have cited some experiments that confirm it. If true of pigs, it is equally true of cattle and sheep. If generally admitted, it will lead to a more liberal system of feeding and to the production of more and far better meat.

It may be thought that I should have said more in regard to the different breeds of pigs in the United States. There is in almost every section a class of useful pigs of more or less local reputation; but it is doubtful if they have been kept pure for a sufficient length of time to warrant us in speaking of them as established breeds. And even if this were the case, I know of none of them that possesses the smallness of offal, perfection of form, early maturity, and fattening qualities of the Yorkshire, Essex or Berkshire. There is none of them that would not be improved in these respects by crossing with a thorough-bred boar of either of these breeds.

Of the diseases of pigs I have said little, for the simple reason that I know little in regard to them. Cleanliness and good treatment are the best medicines for a pig. Anatomically, a pig approximates more closely to a man, than any other of our domestic animals, and if we know how to treat a cold or a diarrhœa in ourselves, we shall not be far wrong in treating a pig in the same way. And so of other diseases. It should be observed, however, that a pig grows as much in eight months, as a man does in eighteen years. This rapid growth enables the pig either to throw off disease in a few days, or failing in this, the disease soon spreads throughout the whole system and carries off its victim. Thus typhoid fever is often so rapidly fatal as to be popularly spoken of as "Hog Cholera." Our first aim, therefore, should be to guard against all hereditary diseases in the selection of pigs for breeding and to exercise great care in maintaining the health and vigor of our swine.

In preparing this book, I have corresponded with many experienced breeders, and in the appendix have given some extracts from this correspondence.

We have been asked by a scientific friend to call this a book on "the Hog" instead of on "the Pig." If it were a work on natural history, hog would be the proper word, but it is purely a practical treatise on domestic swine. A pig is a young hog; and the aim of this work is to induce farmers to so breed and feed their pigs, that they will be in the pork barrel long before they attain the age of an old-fashioned hog. It is proper to speak of "the wild hog," and there may be varieties of swine so little improved as to be hogs still. Let those who have them call them hogs, but we cannot see the propriety of calling a highly refined Essex or Berkshire pig, a hog. All the modern agricultural writers on swine seem to have adopted this view. Not one of them speak of the improved breeds as hogs. Stephens, in his *Book of the Farm*, and the writers in Morton's Cyclopedia of Agriculture, treat of pigs, not hogs. And Youatt, Martin, Richardson, Sydney, and Darwin, all speak of domestic swine as pigs, and it is hardly worth while for us to endeavor to change the usuage of the best writers. We have no desire to have our Western friends speak of the "Magie Hogs" as

Pigs. We presume Hogs is the appropriate name for them; but if they should find it to their interest to cross them with some of the refined thorough-breds, the grades, if well fed, will arrive at maturity before they become hogs. The wants of consumers, and the interests of producers, call for more pigs, and fewer hogs, and it is the object of this work to advocate the change.

J. H.

Moreton Farm, Rochester,
N. Y., April, 1870.

CONTENTS.

CHAPTER I.
Introductory...Page 9

CHAPTER II.
Breeds of Pigs... 14

CHAPTER III.
The Form of a Good Pig... 17

CHAPTER IV.
Desirable Qualities in a Pig... 20

CHAPTER V.
Large vs. Small Breeds and Crosses... 22

CHAPTER VI.
Value of a Thorough-bred Pig... 35

CHAPTER VII.
Good Pigs Need Good Care... 37

CHAPTER VIII.
The Origin and Improvement of our Domestic Pigs................................ 41

CHAPTER IX.
Improvement of the English Breeds of Pigs...................................... 47

CHAPTER X.
The Modern Breeds of English Pigs.. 56

CHAPTER XI.
Breeds of Pigs in the United States.. 98

CHAPTER XII.
Experiments in Pig Feeding 118

CHAPTER XIII.
Lawes and Gilbert's Experiments in Pig Feeding.....................122

CHAPTER XIV.
Sugar as Food for Pigs...135

CHAPTER XV.
The Value of Pig Manure...137

CHAPTER XVI.
Piggeries and Pig Pens..144

CHAPTER XVII.
Swill Barrels, Pig Troughs, Etc..169

CHAPTER XVIII.
Management of Pigs...175

CHAPTER XIX.
English Experience in Pig Feeding..181

CHAPTER XX.
Live and Dead Weight of Pigs...190

CHAPTER XXI.
Breeding and Rearing Pigs...192

CHAPTER XXII.
Management of Thorough-bred Pigs...203

CHAPTER XXIII.
The Profit of Raising Thorough-bred Pigs..................................220

CHAPTER XXIV.
Cooking Food for Pigs...221

CHAPTER XXV.
Summary..230

CHAPTER XXVI.
Appendix..236

HARRIS ON THE PIG.

CHAPTER I.

INTRODUCTORY.

Domestic animals are kept for several objects. The Horse, Mule, and Ass, for labor; the Ox for labor and beef; the Cow for milk and beef; the Sheep for wool and mutton, and in some countries for milk also; Poultry for feathers, eggs, and meat. The Pig, agriculturally, is kept for meat alone. The sole aim of the breeder is to obtain a pig that will produce the largest amount of pork and lard from a given quantity of food.

The same is true of cattle when kept solely for beef. In this case the main difference between the two animals is, that the ox is provided with four stomachs, and is capable of extracting sufficient nutriment, in ordinary cases, from bulky food, while the pig has but one stomach—and that comparatively a small one—and, consequently, requires food containing a greater amount of nutriment in a given bulk. Grass is the natural food of the ox; roots, nuts, and acorns, worms and other animal matter, the natural food of the hog. The pig unquestionably requires a more concentrated food than the ox or the sheep.

The stomach of an ox weighs about 35 lbs.; that of a Southdown or Leicester sheep from 3 to 4 lbs.; and that of a pig $1\frac{1}{4}$ lbs.

The weight of the stomach, in proportion to each one

hundred pounds of live weight, is: ox, 3 lbs.; sheep, 3 to 4 lbs.; fat pig, 0.66 lbs. In other words, in proportion to live weight, the stomach of an ox, or sheep, is about five times as great as that of a pig.

It is quite evident, from these facts, that the pig is not so well adapted to feed on grass or hay as the ox or sheep.

This is a strong argument *against* the hog as an economical farm animal.

In proportion to the nutriment they contain, the concentrated foods are more costly than those of greater bulk. Not only is their market price usually higher, but it costs more to produce them. Elaboration is an expensive process. The common white turnip, containing from 92 to 94 per cent of water, can be grown with less labor and manure, and in a shorter period, than the Swedish turnip, containing from 88 to 90 per cent of water, and this less than the Mangel Wurzel, containing only 86 per cent of water. Carrots, which are still more nutritious, are even more costly, in proportion to the nutriment they contain. This is probably a general law.

As the ox can subsist and fatten on less cencentrated and less costly food than the pig, it follows, therefore, that a pound of beef ought to be produced at less cost than a pound of pork.

There are, however, several circumstances which modify this conclusion. Pigs will eat food which, but for them, would be wasted. Where grain or oil-cake is fed to cattle, a certain number of pigs can be kept at a merely nominal cost. We can in no other way utilize the refuse from the house and the dairy so advantageously as by feeding it to swine. On grain farms, pigs will obtain a good living for several weeks after harvest, on the stubbles, and in some sections, they find a considerable amount of food in the woods.

Even where we have none of these advantages, the difference in the cost of producing a pound of beef and a pound of pork is not so great as the above considerations would lead us to suppose. The hog is a great eater. He can eat, and digest, and assimilate, more nutriment in a given time, in proportion to his size, than any other of our domestic animals.

The extensive and elaborate experiments of Messrs. Lawes and Gilbert show that, notwithstanding pigs are fed much richer food than oxen and sheep, they nevertheless eat about twice as much food, in proportion to live weight, as a sheep. On the other hand, it was found that 401 lbs. of Indian corn meal and bran (dry) produced 100 lbs. of pork (live weight), while it required 1,548 lbs. of oil-cake and clover hay (dry) to produce 100 lbs. of mutton (live weight.)

Why a pig should gain so much more from a given quantity of food, than a well-bred sheep or steer, has not hitherto been explained. It has been attributed to the fact that the pig possesses larger and more powerful assimilating organs.

Thus, Messrs. Lawes and Gilbert say: "An examination of these tables [of results of experiments] will show that the stomachs and contents constituted

In the oxen about 11½ per cent of the entire weight of the body.
" " sheep " 7½ " " " " " " "
" " pig " 1¼ " " " " " " "

"The *intestines* and their contents, on the other hand, stand in an opposite relation. Thus, of the entire body, these amounted

In the pig to about 6¼ per cent.
" " sheep " " 3½ " "
" " oxen " " 2¾ " "

"These facts," they remark, "are of considerable interest, when it is borne in mind, that in the food of the ruminant there is so large a proportion of indigestible woody fibre, and in that of a well-fed pig a comparatively large proportion of starch—the primary transformations of which are supposed to take place chiefly after leaving the stomach, and more or less throughout the intestinal canal."

These facts explain very clearly why an ox or a sheep can thrive on more bulky food than a pig; also why a pig can assimilate more food than an ox or a sheep, but they do not show why *a given amount of food* should produce so much more flesh and fat when fed to the pig than when fed to oxen or sheep—unless, indeed, we are to suppose that in the case of the ox and the sheep, a considerable proportion of the food passes through the body undigested and unassimilated. But an analysis of the excrements indicates nothing of this kind. Except when an excessive amount of grain is allowed, the food is unquestionably as thoroughly digested and assimilated in the ox and the sheep, as in the pig.

We must, therefore, look for some other explanation of the fact that pigs can gain more rapidly on a given amount of nutriment than oxen or sheep.

An animal requires a certain amount of nutritive matter merely to sustain life. This matter may be derived either from the daily food supplied, or from matter previously stored up in the body. The actual amount required, varies greatly according to the conditions in which the animal is placed. If kept comfortably warm and quiet, less is required than if exposed to cold, or compelled to labor. But in all cases, wherever life exists, a certain amount of nutritive matter is necessary for its support. Directly or indirectly, this is always derived from the food.

How much food is necessary to keep an animal so that it shall neither gain nor lose in flesh, has not been accurately ascertained. Thousands of animals are so kept, but the actual amount consumed is seldom determined. It often happens that cows, not giving milk, are so kept during the winter that they do not weigh a pound more in the spring than in the fall. We receive absolutely nothing for the food they eat. It is all consumed in sustaining the vital functions.

A well-bred Shorthorn has been made to weigh 1,200 lbs. by the time it was a year old. On the other hand, an ox is sometimes kept five years before it attains this weight. The Shorthorn was fed a considerable amount of food over and above that required to sustain life, while the other had little more than was necessary for this purpose. Let us assume that the latter ate 4 tons of hay a year, and that 80 per cent of it was used merely to sustain life. At the end of five years he would have consumed 20 tons of hay, 16 tons of which have been used merely to sustain the vital functions, and 4 tons have been converted into 1,200 lbs. of animal matter.

The Shorthorn accomplished the same result in one year; and we may reasonably suppose that in this case also, 4 tons of hay or its equivalent, were sufficient to furnish the material necessary for the formation of this amount of animal growth. We may further assume that at any rate no more food was required to sustain the vital functions in the Shorthorn than was required by the other animal. This we have estimated at 3 tons 4 cwt. a year. It follows, therefore, that the Shorthorn, by eating 7 tons 4 cwt. of hay or its equivalent, in a single year was enabled to produce as much beef as the other steer produced by the consumption of 4 tons a year for five years. The consumption of less than twice as much food enabled the Shorthorn to increase five times as rapidly as the other. Seven tons 4 cwt. of hay, or its equivalent, produced as much growth (and probably more beef and fat), when fed to the animal capable of eating and assimilating it.

These considerations will show why a pig, that can eat so much more food than a sheep or an ox, in proportion to size, is enabled to grow so much faster, in proportion to the food consumed.

The fact that the pig has greater powers of assimilating

food, merely explains why he can grow so rapidly, but it throws no light on the fact that he can gain more rapidly, in proportion to the food consumed, than any other domestic animal. The real explanation of this fact is the one given above. He can eat more, digest more, and assimilate more, over and above the amount of food necessary to sustain life.

CHAPTER II.

BREEDS OF PIGS.

Like all other animals, pigs adapt themselves to the circumstances in which they are placed. Where the supply of food is scanty and uncertain, they grow slowly, and are long in coming to maturity. Where they have to travel far in search of their food, they have legs adapted for the purpose; and if they are obliged to seek their food under ground, their snouts soon become long and powerful. Where they are liable to molestation or attack, they soon acquire a ferocious disposition and the means for defence. On the other hand, where they have a liberal and constant supply of food, where they are provided with warm and comfortable quarters, and are never harshly treated, they become gentle in disposition, are indisposed to roam about, have finer hair and skin, shorter and finer legs, smaller head, ears and snout. They grow rapidly and mature early.

Such a change does not take place at once; and the same may be said of the conditions. A rude system of agriculture is never immediately followed by high farming. There must be intermediate changes. And so it is with our domestic animals. We have almost as many kinds of hogs as we have different kinds or systems of farming. We do not call them *breeds*, because there is little permanency of character about them. They are constantly changing, just as the management of their owners varies.

A breed possesses fixed characteristics. If fully established, and the conditions of feeding and management are not changed, these characteristics are transmitted from generation to generation. In pigs, owing to their fecundity, it is a comparatively easy matter to establish a breed.

Man does not create a breed. God alone creates. All that we can do is to avail ourselves of that inherent disposition which animals have of adapting themselves to the conditions in which they are placed. The conditions are under our control. Let the breeder first make up his mind what system of feeding and management he will adopt. Then let him steadily and perseveringly adhere to it. An unstable man can never be a successful breeder. If he wishes a breed that will grow moderately on a moderate allowance of food, and arrive at maturity in two or three years, he can attain his object by feeding moderately and selecting such pigs to breed from, as come nearest his wishes. If any pigs in the litter manifest a disposition to grow rapidly, they must be rejected. Such pigs are not suited to a moderate allowance of food. Their offspring will certainly degenerate. Better select those which make the slowest growth, and which are consequently least likely to experience the injurious effects of starvation. By steadily pursuing this method, a breed can be obtained which will eat little and grow slowly, and yet remain healthy. If it is desired to have them attain a greater weight without increasing the daily allowance of food, attention must be directed to this object. Do not let either the sow or the boar breed until they have attained their fullest growth, say at three, four, or five years of age.

The advantage of such a breed lies in the fact that it would suffer less from occasional starvation, than breeds which are adapted to grow rapidly, and mature early, on liberal feeding. But of course such a breed can only be profitable where the food costs little or nothing—and even in this case it may well be questioned whether a breed that eats more and gains faster would not be more profitable. All that we wish to show is, that no matter what the object of the breeder is, he can attain it. He can raise a breed adapted to any system of feeding and management he desires to adopt. In point of fact, the pigs will adapt themselves, sooner or later, to the supply of food and the means necessary for them to use, in order to obtain it. The breeder can, by selection, greatly accelerate the change, but the main cause is the food and treatment. In this sense the "breed goes in at the mouth."

If a farmer wishes a breed of pigs that will grow with great rapidity and fatten early, he cannot attain his object without liberal feeding. If he will furnish this for several generations and at the same time provide warm and comfortable quarters, and never suffer the pigs to be harshly treated or neglected, he will do much to secure his object. Selection will do the rest. It is generally supposed that the success of the breeder depends mainly on his ability to select a boar—having those points fully developed in which his sows are most deficient; and doubtless this requires much skill and nice discrimination. But we are satisfied that the cause of failure is generally owing to inconstant or illiberal feeding. The breeder must love his animals, and must give them his constant personal attention. A few weeks' neglect, starving at one season and surfeiting at another, harsh treatment, and damp, dirty pens, will counteract all the advantage derived from months of good management.

Nature protects herself. The offspring of animals liable to such occasional neglect will, so to speak, *expect* such treatment, and even if they themselves have liberal and constant feeding, they will not possess the qualities of rapid growth and early maturity, in the highest degree.

It is the weakest link that determines the strength of a chain. And so far as inherited qualities are concerned, the rapidity of growth will be influenced more by the periods of neglect and starvation, than by the occasional periods of high feeding. Starving a young, well-bred sow may not show any great and injurious effect on the sow herself, but the offspring of such a sow, if she breed at all, will be seriously injured. A few months starvation and neglect may counteract nearly all the advantages which the breed has acquired by generations of careful breeding and feeding.

CHAPTER III.

THE FORM OF A GOOD PIG.

The aim of all breeders of animals designed solely for meat, is to have the body approximate as closely as possible to the form of a parallelopiped. In proportion to the size, an animal of this form contains the greatest weight. Hence it is, that farmers who have kept nothing but common pigs, and who look upon a well-formed, grade Essex or Suffolk as "small," are surprised to find, when brought to the scales, that it weighs more than an old-fashioned, ill-formed pig of much greater apparent size.

Another advantage of this form is, that it gives a greater proportion of the most desirable parts of the pig.

In a pig of this form the ribs are well-arched. We cannot have a flat, broad, "table-back" without this. And consequently the muscle which runs along each side of the vertebræ, is well developed, and we have a large quantity of meat of the best quality.

This form also affords abundant room for the lungs, stomach, and intestines; and it is on the capacity of these organs to convert a large amount of comparatively cheap food into a large quantity of flesh and fat that determines the value of the animal.

We annex a portrait of a tolerably well-formed pig, with lines showing how to apply the test above alluded to. The nearer he will fill the rectangular frame, the nearer he approaches to perfection of form. It would be well, for farmers to place a straight cane along the back, also along the sides, shoulders and hams of their pigs, and see how near they come up to the desired standard.

Fig. 1.—TESTING THE FORM OF A PIG.

The length of a pig should bear a certain proportion to his breadth. Many farmers object to the improved breeds, because they are too short. In point of fact, however, they are often longer than their ill-bred favorites. They appear short, because they are so broad. A large-boned hog is longer than one having small bones. There are as many vertebræ in the shortest Suffolk as in the longest Yorkshire.

A fine-boned pig cannot be long-bodied. It may appear long, but this will usually be because it is narrow. Breadth and depth are of far greater importance than length. Robert Bakewell, the originator of the improved Leicester sheep, and one of the most skillful and experienced breeders in the world, is said to have formed a breed of pigs that, when fat, were "nearly equal in height, length, and thickness, their bellies almost touching the ground, the eyes being deep set and sunk from fat, and the whole carcass appearing to be a solid mass of flesh." Bakewell left no record of his mode or principles of breeding, but the following sentence from the description of his pigs above quoted, throws light on the point we are now considering: "These pigs are remarkably fine-boned and delicate, and are said to lay on a larger quantity of meat, in proportion to bone and offal, than any other kind known." In other words, Bakewell, with all his skill, could not obtain fineness of bone, and length too, any more than a builder could reduce the size of his bricks, and then make the same number form as long a wall. What he probably did, was, to take a large pig and reduce the size of the bones, and consequently the length of body, without reducing the breadth and depth of the animal.

In a common sow, to be crossed with a thorough-bred boar, length of body is often very desirable; but in a thorough-bred pig it is a doubtful quality, as indicating a want of breadth and fineness of bone.

The head of a pig should be set close to the shoulders. The broader and deeper the cheeks, the better, as next to the ham and shoulder there is no choicer meat on the pig. A well-cooked cheek of bacon, with roast chicken, is a dish for an epicure.

The snout should be short and delicate, and the ears small and fine. A thick, heavy, pendant ear is an indication of coarseness and is never desirable in a thorough-bred pig. It should be small, fine, soft, and silky. It should be well set on the head and lean a little forward, but not fall over. An ear that is upright indicates an unquiet disposition.

CHAPTER IV.

DESIRABLE QUALITIES IN A PIG.

As the domestic hog is kept solely for its flesh and fat, the pig that will afford the greatest amount of meat and lard of the best quality at the least cost, other things being equal, is the most profitable breed.

It has been well said that Cincinnati owes its wealth to the discovery of a method of putting 15 bushels of corn into a three-bushel barrel, and transporting it to distant markets. This has been accomplished by means of the pig. He converts 7 bushels of corn into 100 lbs. of pork.

In accomplishing this result, the organ of first importance is the stomach. It is here that the first change in this wonderful process commences. In a flouring mill we have a water-wheel or steam-engine which drives the stones, and the machinery for removing the bran and other inferior products of the grain from the fine flour. The capacity of the establishment is determined by the motive power and the "run of stones." A pig is a mill for converting corn into pork. The stomach is at once the water-wheel or steam-engine, and the stones for grinding the grain,—and the motive power, which runs the mill and the machinery, is derived from the consumption of corn.

Now, if we furnish merely corn enough to run the machinery, and put no grain in the hopper, we lose not only the use of the mill, but of all the grain used for fuel.

If we should keep the mill supplied only half the time, and yet keep the machinery running at full speed night and day, (as we must needs do in the case of an animal) would it be considered good management?

Let us see. Suppose it takes 75 lbs. of corn to run the machinery. If we furnish no more than this, we get nothing in return. If we furnish 100 lbs., (say 75 lbs. for fuel and 25 lbs. for the hopper,) we may obtain, say 20

DESIRABLE QUALITIES IN A PIG.

lbs. of flour. If we furnish another extra 25 lbs. to the hopper, or 125 lbs. in all, we get 40 lbs. of flour; if we furnish 150 lbs., we get 60 lbs. of flour. In other words, 150 lbs. of corn will furnish three times as much flour as 100 lbs.

It may be said that more power would be required to run the mill when it is grinding than when it is running empty. But in the case of an animal it is doubtful how far this objection holds. It is not improbable that the conversion of each additional pound of corn into pork generates the amount of power necessary for the change. But whether this be so or not, no one can question the advantage to be derived from furnishing all the grain that the mill will grind and manufacture.

Of the desirable qualities in a pig, therefore, a vigorous appetite is of the first importance. A hog that will not eat, is of no more use than a mill that will not grind. And it is undoubtedly true that the more a pig will eat in proportion to its size, provided he can digest and assimilate it, the more profitable he will prove.

The next desirable quality is, perhaps, quietness of disposition. The blood is derived from the food, and flesh is derived from the blood. Animal force is derived from the transformation of flesh. The more of this is used in unnecessary motions, the greater the demand on the stomach, and the more food will there be required merely to sustain the vital functions—and the more frequently flesh is transformed and formed again, the tougher and less palatable it becomes.

This quality, or quietness of disposition, combined with a small amount of useless parts or offal, has been the aim of all modern breeders. Its importance will readily be perceived if we assume that 75 per cent of the food is ordinarily consumed to support the vital functions, and that the slight additional demand of only one-sixth more food, is required for the extra offal parts and unnecessary activity. Such a coarse, restless animal would gain, in flesh and fat, in proportion to the food consumed, *only half as fast* as the quiet, refined animal.

A little calculation will show this to be true in theory, as it is undoubtedly true in practice. Thus take two pigs. No. 1 eats 100 lbs. of corn, 75 lbs. of which are required to sustain the vital functions. He gains, say 20 lbs.

No. 2, a coarse, restless pig, eats 100 lbs. of corn, $87\frac{1}{2}$ lbs. of which are necessary to support the vital functions.

No. 1 has 25 lbs. of food over and above the amount required to sustain the vital functions, and gains 20 lbs. of pork. No. 2 has only $12\frac{1}{2}$ lbs., and consequently, cannot produce more than 10 lbs.

To assume that a rough, coarse, savage, ill-bred, squealing, mongrel hog will require only one-sixth more food to "run his machinery," than a quiet, refined, well-bred Berkshire, Essex or Suffolk pig will not be considered extravagant; and yet it undoubtedly follows that, for the food consumed, the quiet pig will gain in flesh and fat twice as fast as the other. If in addition to this he will eat 25 per cent more food, he will gain *four times* as fast.

The two great aims of every pig breeder should be to lessen the demands on the stomach for offal or least valuable parts, and for unnecessary activity on the one hand, and on the other to increase the power of the stomach, and digestive and assimilative organs as much as possible.

CHAPTER V.

LARGE vs. SMALL BREEDS AND CROSSES.

Mr. Lawes' experiments on the different breeds of sheep, prove conclusively that well-bred mutton sheep of the same age, consume food in almost exact proportion to their size or live weight. Two Cotswold sheep, weighing 120 lbs. each, will eat as much food as three Southdown sheep, weighing 80 lbs. each. But the two Cotswolds will gain much more than the *three* Southdowns. The average increase for one hundred lbs. live weight was, with the Cotswold, 2 lbs. 2 oz. per week; and with the Southdowns, 1 lb. 10¾ oz. per week—both breeds having precisely the same food. In other words, two Cotswold sheep, weighing 120 lbs. each, would eat the same amount of food as three Southdowns weighing 80 lbs. each; but the two Cotswolds would gain 17 lbs. each, while the three Southdowns gained only 9 lbs. each. Where Cotswold mutton brings as much per pound as the Southdown, it is evident that the Cotswolds are the more profitable breed for fattening.

We know of no similar experiments on the different breeds of pigs. Reasoning from analogy, we might conclude that, as the large Cotswold sheep gained much more, for the food consumed, than the small Southdowns, the large Yorkshire pigs would gain much more, for the food consumed, than the small Suffolks.

This may or may not be true. If it should prove to be a fact, we should conclude that a pig of the large breed ate much more food over and above the amount required to keep up the animal heat and sustain the vital functions, than a pig of the small breed; and, as we have attempted to show in a previous chapter, the large pig would, in such a case, gain much more in proportion to the food consumed, than the small pig of the same age.

There can be no doubt that a large pig, other things being equal, will eat more food than a small pig of the same age.

It is equally true that a large pig, at ordinary temperatures, will not require, in proportion to its weight, as much food to keep up the animal heat as a small pig. A pig weighing 100 lbs. will not radiate as much heat as two pigs weighing 50 lbs. each. The larger the pig, the less surface is there exposed to the atmosphere in proportion to weight.

It follows, therefore, that a large pig, eating more food and losing less animal heat, would have a greater amount of food to be appropriated to the formation of fat and flesh, in proportion to live weight, than a smaller pig of the same age.

So far as this kind of reasoning goes, therefore, it would seem that the large breeds of pigs are preferable to the small breeds.

This conclusion is opposed to the opinion of a large number of very intelligent and observing pig breeders and feeders. There can be no doubt that the weight of testimony, so far as the production of a given amount of pork from a given amount of food is concerned, is against the large breeds.

The truth of the matter is probably this: The small breeds mature earlier than the large breeds. This in itself is a great advantage. The pigs are not only ready for the butcher at an earlier age, but as animal life is always attended by a constant transformation of tissue, every day we gain in time, saves the amount of food necessary to supply this waste and keep up the animal heat.

Early maturity, therefore, is one of the principal aims of the breeder and feeder. But early maturity is always attended with a diminution of size; and the small breeds owe their value, not to their small size, but to their early maturity and tendency to fatten while young.

In point of fact, however, the term Small Breed or Large Breed, as used by our Agricultural Societies, has no very distinct meaning. The New York State Agricultural Society offers prizes for two classes of pigs—and only two.

1st. "Large Breed; which, when full grown and fattened, will weigh over 450 lbs. dressed."

2d. "Small Breed; which, when full grown and fattened, will not weigh over 450 lbs. dressed."

Exhibitors seem to have entered their pigs in the class for small breeds one year, and in that for large the next. Berkshire, Essex, Suffolk and Yorkshire have all been exhibited, first in one class and then in another, and frequently the same breeder will exhibit Berkshire or Essex at the same fair in both classes.

The same state of facts seems to exist in England. There are Large Yorkshires and Small Yorkshires, Large Berkshires and Small Berkshires. Of late years, a new class of "Medium" Breeds has been formed at the Agricultural Shows. There, as here, it is not always easy to determine the class to which a particular breed belongs. An English breeder of "*Small* Yorks," says he can "get them up profitably to 600 lbs. when thick bacon is required."

On the other hand, the advocates of the *Large* Yorkshires claim that pigs of this breed "attain a good bacon size at a very early age, and when killed, they cut more lean meat in proportion to the fat than the smaller breeds."

A sow of this breed, which took the Prize at Rotherham, in 1856, age three years and two months, weighed 1,315 lbs.*

The author above quoted, says: "The large breed is equally valuable for making large or small bacon, that being only a matter of age; as porkers of a few weeks old, they are unequaled; their flesh being very rich and well-flavored, and not so fat as the small breeds."

On the other hand, Mr. George Mangles, of Givendale, Ripon, one of the largest and most successful breeders and feeders in Yorkshire, furnishes the London Farmers' Magazine, for June, 1861, the following interesting account of his experience:

"About ten years ago, I commenced pig-keeping on a

---

* Youatt on the Pig. By S. Sidney. London: 1860. Page 14.

larger scale than the generality of farmers. What I wanted, and what my farm required, was a quantity of good manure. I first tried buying stores in the neighborhood, but soon gave that up, as they were chiefly of the large breed, and required too much food and liberty. I had no alternative but to breed my own stores. With a view to find a profitable sort, I purchased a few of the best from different breeders of note, and kept them separate, and also a few stores of each sort together, living on the same kind of food. I also tried the different crosses; but, to get the cross, I must have pure stock at first; so I considered it best to keep to a pure breed. I tried the Essex, the black Leicester, the Berkshire, the large Yorkshire, the small Yorkshire, and lastly the Cumberland small breed. I must confess that at the outset I had but little experience to guide me; not understanding the principles of breeding, I committed many foolish mistakes, which I paid dearly enough for; and if these few lines should meet the eye of any one wishful to form and keep a breed of pigs, I shall be glad for such a man to profit by the experience of another. I never expected pigs to live on nothing: because the manure made from pigs living on nothing would be worth nothing, and it was good manure I was aiming at. I found any breed pay, except the large breeds. All the crosses having the small breed for the sire always paid: whichever breed is intended to be kept, the best bred ones should be obtained. I do not advocate breeding in-and-in; but I do advocate, if you want to maintain the same style of animal, generation after generation, to cross with the same blood, but as far distant as you can get it. I do not know a better sign of pure breeding than a litter of pigs all alike, or three or four sisters breeding alike to the same boar. When the breed is obtained, one thing must always be kept in mind, the first boar a sow is put to, influences the succeeding litters for three or four times.

"After the breeding, come the feeding and attention. Milk and fat must go in at the mouth before it makes its appearance in the animal. I do not believe those, who say their pigs get fat on nothing. I know from experience that one pig would live where another would starve, and what it would take to make one large-bred pig fat, would make several smaller-bred ones 'up.' A great help to profitable pig-keeping is warmth, and confinement, and regularity in feeding; as by also keeping the skin of the animal clean by washing and brushing occasionally. If two animals of the same litter be put into two different sties, and have the same quantity of food each, the one that is kept warm and with the skin clean will gain more weight than the other. I found that out one winter, when Jack Frost was astir, before I put up a new pig-shed. My man was feeding a lot of pigs alike, only some were in common sties and others in a warm shed. The difference was very striking: those kept warm fed nearly half as fast again as the others. This induced me to build a long covered shed sixty feet long and eighteen feet wide, that would hold seventy porkers or fifty bacon pigs, where, when the thermometer has been below freezing point outside, it has inside been very warm and comfortable. The pigs have their food warm in winter, and are never starved by the cold; they are bedded with clean straw every other day, and the shed is kept rather dark. The manure made is of first quality and fit to use for turnips.

"Perhaps some of the readers of this paper would like to know something about the dietary of my pigs. I have not included sugar in my list of feeding ingredients. I have never gone higher than new milk, which they always take without sweetening. In the first place, I must say that I exhibit at a few of the leading agricultural meetings, and am generally, if not at the top of the ladder, not many spokes off. I keep my breeding stock different to my show stock, as I do not like breeding animals to be over-fat; but show animals are obliged to be fat, or the judges will pass them over. The over-feeding of prize animals is a very great evil, but one that can not be very well remedied. A show of lean breeding animals would be a very *lean* show indeed in many respects: an exhibitor must always sacrifice some of his best animals to please the public fancy. I think there is less risk in fat breeding pigs than any other animal. I have had several very fat sows pig, and never lost any. I gave them nothing but a very little bran and water a week before pigging, and but little after for a week, while I put a little castor oil in their food directly after pigging. I have the greatest trouble in reducing the male animals, as they will nearly hunger to death before they will part with their fat. I generally turn them into a large yard, and give them plenty of water, and a wurzel or two every day, or turn them out to grass in summer.

"To my regular breeding pigs and stores, I am giving boiled rape-cake and barley-meal, one feed a day, and one feed of raw potatoes or wurzel; and if in summer, I turn them to grass, or soil them with clover in the yards.

"I soil a good many every year. A week or two before the sow pigs, I contrive to put her into a loose box, with a railing around to keep her from crushing the pigs. I can always tell when she is going to pig by trying if she has milk in the paps: if a sow gives milk freely, she will pig any time. I then contrive to be, or have some one, near at hand, to take the pigs away as she pigs them, as the sows are sometimes uneasy and will crush them. After she has pigged, I feed her with warm water and bran, and then give her the pigs and leave them, because the less they are disturbed the better. I always feed the sow sparingly at first, as I have sometimes found, when a sow has been fed too liberally at first, the flow of milk is greater than the pigs can take; consequently the udder becomes hard, and the sow is very uneasy, and will scarcely let the pigs suck her. If such is the case, the best way is to rub the udder well with the hand three or four times a day. Small-bred sows are commonly very quiet and tractable.

"Generally when the pigs are three weeks or a month old, they will scour, if proper care has not been paid to the sow's feeding. I never could get a man that could get me a litter through without scouring. I have tried different plans, but the one I have found most successful is, to always give the sow a tablespoonful of the following mixture in her food: Mix together 2 lbs. of fenugreek, 2 lbs. of anise-seed, ½ lb. of gentain, 2 oz. carbonate of soda, and 2 lbs. of powdered chalk. The sow gets very fond of this, and the little pigs, too, like it. Give the pigs also plenty of coal ashes to root amongst. I prefer oats, wheat, and a little barley ground together, for sows giving milk. I have never tried the sugar diet, but I have found new milk fresh from the cow to work wonders in a short time.

"Warmth, cleanliness, and regularity in feeding, a little good food and often, are the main secrets in rearing young pigs. I never like to see food left in a pig's trough: just give what they can eat up and no more. When pigs are put up to feed they should be kept warm and quiet. Five porkers or three bacon pigs are plenty together. The pen they are kept in need not be very large, but the pigs should be rung, and a little fresh bedding spread about them every second day. Pigs like to be kept warm, but plenty of fresh air must be allowed to circulate through the pens, or else disease will soon show itself."

According to the editor of Youatt on the Pig, Mr. Mangles "is a plain farmer, feeding pigs for profit," and his statements will be received with all the more confidence on this account. We give the details of his management, not only because they are interesting and instructive in themselves, but because the system of management and feeding have often more to do with the profits of pig breeding and feeding than the mere question of large or small breeds.

On page 66 we give a portrait of one of his Prize pigs of the Small Breed, from a steel engraving in the London Farmers' Magazine for June, 1861.

It will be observed that Mr. Mangles says he "found any breed pay except the large breed." "All the crosses having the small breed for the sire always paid."

To the same effect is the testimony of Mr. Hewitt Davis, a name familiar to all readers of English agricultural literature. He says:

"My experience in stock keeping has been so decidedly in favor of breeding and fatting of pigs, that I may, with advantage to many who think differently, give some account of my management. That I should do so is the more necessary from farmers having generally a very low opinion of the profit to be gained from the breeding of pigs, and I cannot but ascribe their failures too often to the negligence with which this stock is looked after. On an arable farm of 200 acres my stock has been 12 sows and two boars; and their produce, according to the season, consisted either of rising stores running in the yards, or on the leas or stubbles; or of porkers in the sties fatting for the market. From March to October my stock may be said to have lived loose on store keep, principally green food; and from October to March (the parent stock excepted) in sties, fatting on roots and boiled corn. The sows on an average gave me, one with another, 14 pigs a year each, so that in summer my stock was about 100 upon store keep, and in winter about 200, of which 180 were in sties finishing for market. The spring litters went off in January and February as large porkers of 30 stones (240 lbs.) each, and the autumn-born as small porkers of about 7 stones (56 lbs.) each; the first realizing about £5 each, and the last about 30s. each, so that each sow returned about £45 a year; and this amount there is no difficulty in obtaining, large pork selling at 3s. 4d. per stone of 8 lbs. (11 cts. per lb.), and small pork at 4s. 4d. (14 cts. per lb.). Success, in the raising of pig stock, I found, was to be attained only by attention to fully carrying out the following principles—viz., the accommodation for pigs must be sunny, dry, sheltered from cold wind, and yet well ventilated. Their sties being carefully protected on the north, east, and west sides, and open only on the south; so that whilst no cold winds can have access, there should be no obstruction to the sun shining in and on to their beds. The pigs must be regularly and carefully attended; sufficient should be kept to wholly occupy their attendant's time, and to them should that attendant's time and attention be wholly given. An old man is better than a young one; and this is an office suited to one infirm or past general labor. The sows must never be permitted to farrow earlier than the end of March, nor later than October. The cold of winter is fatal to many farrows, and young pigs are ill able to bear up against it. Provide roots (potatoes, kohlrabi, swedes, carrots, and mangel wurzel) for their keep, aided with boiled corn, from September to June; and tares, clover, beans, and maize, green, from May to September. Breed from large, strong sows, with boars of the finer breed, having in view the gaining of large farrows, good nursing, and a rapid attainment of weight; look to the mother for nursing, and the father to correct coarseness of form in the mother. Attached to the sties have a boiling-house with copper and food cisterns; and in front of the sties a yard for the pigs to be turned into. Attention to these points makes all the difference between profit and loss."

The point in Mr. Davis' statement to which we wish to call particular attention is this: "Breed from large, strong sows, with boars of the finer breed, having in view the gaining of large farrows, good nursing, and a rapid attainment of weight; look to the mother for nursing, and the father to correct coarseness of form in the mother."

In other words, aim to get the digestive powers of the large breed in the body of a small, highly refined pig. Increase the supply of food and lessen the demand upon it for everything except the formation of flesh and fat.

It will be found that, consciously or unconsciously, all the eminently successful pig feeders have aimed to attain this result.

The question of Large vs. Small Breeds, therefore, can only be answered by taking these objects into consideration. We need both breeds. The large breed to give us sows, and the small breed to give us boars. It is a mistake to refine and reduce the size of the large breed, and then to breed from these "improved" pigs of the large breed. To produce pigs merely for the butcher, we should resort to crosses with a large, vigorous, unpampered sow put to the finest, thorough-bred boar of the small breeds that can be obtained. The larger the sow and the smaller the boar, the more will the little pigs be able to eat in proportion to their size, and the greater will be their growth in proportion to the food consumed.

Mr. John Coate, a breeder of "Improved Dorsets," who took the Gold Medal, five years in succession, at the Smithfield Club Show, for the best pair of pigs, says:

"Crosses answer well for profit to the dairyman, as you get more constitution and *quicker growth.*"

One of the most extensive farmers in West Norfolk writes to Mr. Sidney: "The cross between the Berks boar and Norfolk sow (white), *like all cross breeds*, is most profitable to the feeder, but we must have pure breeds first." And Mr. S. adds: "This Norfolk opinion is confirmed by all my correspondents. The Berkshire pig is in favor in every dairy district, either pure or as a cross, *but chiefly as a cross.*"

Again, the same author says: "The Improved Essex is

one of the best pigs of the small black breeds, well calculated for producing pork and hams of the finest quality for fashionable markets; but its greatest value is as a cross for giving quality and maturity to black pigs of a coarser, hardier kind. It occupies, with respect to the black breeds, the same position that the small Cumberland-Yorks do as to the white breeds—that is to say, an improved Essex boar is sure to improve the produce of any large dark sow."

Again: "The Berkshire breed have benefited much from the improved Essex cross. The best Devonshire pigs have a large infusion of this strain. The improved Dorsets, the most successful black pigs ever shown at the Smithfield Club Shows, have borrowed their heads, at least from the Boxted [Essex] breed."

A Bedfordshire farmer writes: "The Woburn breed described by Youatt was a good sort of pig, of no particular character, except great aptitude to fatten. They were discontinued in consequence of the sows being very bad sucklers, in favor of a cross-bred animal, the produce of Berkshire sows and white Suffolk boars, the best that could be got. These are prolific, of good quality, can be fed at any age and to a fair medium weight. A cross like this pays the farmer best."

Mr. Thomas Wright says, the cross of the Berkshire with the Tamworth "produces the most profitable bacon pigs in the kingdom, the Berkshire blood giving an extraordinary tendency to feed, and securing the early maturity in which alone the Tamworth breed is deficient. The cross of the Berkshire boar with large white sows has been found to produce most satisfactory results to plain farmers."

The editor of the work from which these extracts are made says, that the current of opinion among English farmers, both as regards sheep and pigs, is towards crosses. "Breeding pure-bred stock pays well as a separate business, if judiciously conducted; but the ordinary tenant farmer will generally find that a cross-bred sheep, a cross-bred pig, and even a cross-bred ox, *in the first cross*, fattens more profitably than a pure-bred animal."

That this is the general opinion among practical farmers there can be no doubt. But there is no advantage in crossing merely for the sake of crossing. There should be an object in view. We should aim to improve the form, early maturity and fattening qualities of the offspring. In doing this, the tendency always is towards reducing the size. Bakewell reduced the size of the Leicester sheep, and Ellman of the Southdowns. Fisher Hobbs reduced the size of the original Essex pigs by using Lord Western's Neapolitan-Essex boars on selected Essex sows of large size, with good constitutions, and enormous eaters. The Berkshire pig was originally "a much larger and coarser animal than now." The small Leicesters were "the great improvers of the gigantic Yorks."

"What, then," it may be asked, "have we gained by the improvement?"—We have gained this: While the size to which the animal would attain at maturity has been reduced, yet we can get a much greater weight, with less offal, in a given time, and with a far less consumption of food. An improved Essex pig at three years old will not weigh as much as the original unimproved pig at the same age, and with the same food. But at one year old the improved Essex can be made to weigh as much as the other would at eighteen months or two years. They have, *or ought to have*, the digestive powers of the large, old breed, combined with the small bones, little offal, early maturity, and fattening qualities of the Neapolitan Essex. They can eat a large quantity of food, and convert it rapidly into pork of the highest quality.

We say they *ought to* be great eaters, and have powerful digestive organs. But the high feeding necessary to develop the fattening qualities in a breed, is apt to weaken the digestive organs; and it is best, in raising pigs for the butchers, to breed from large, healthy, vigorous sows, and a thorough-bred highly refined boar of a small breed. Such a cross will furnish grades that will eat more and fatten more rapidly than the thorough-breds.

To cross thorough-breds is absurd. There is nothing to be gained by it that cannot be obtained by breeding from common or grade sows with a thorough-bred boar; besides thorough-breds are always more costly than common stock or grades. That a cross, for instance, between a thorough-bred, highly refined Essex boar and a thorough-bred Berkshire sow would afford healthier, hardier, and more profitable pigs for the butcher than either thorough-bred Essex or thorough-bred Berkshires, may be true. It is not an easy matter to maintain the health and high character of any of our improved breeds. In-and-in-breeding, especially with pigs, leads to degeneracy; and all pig breeders find it necessary to introduce a new strain of blood, either from animals bred distinct on their own farm, or, what is considered better, from the same breed kept in another section of country. By judicious selection, in this way, the breed can be maintained or improved. For the same reason, a cross between two distinct breeds, may give a litter of pigs better than either of the parents. But this is not only an expensive way of raising pigs for the butcher, but equally good, if not better pigs can be obtained by using a thorough-bred boar on grade, or common sows, selected with judgment.

CHAPTER VI.

THE VALUE OF A THOROUGH-BRED PIG.

It cannot be denied that many farmers in the United States have purchased thorough-bred pigs, and after keeping them a few years, have given them up in disgust. One cause of this result may be found in the erroneous ideas prevalent in regard to the object of keeping improved thorough-bred animals. No farmer could afford to keep a herd of high-bred Duchess Shorthorns simply for the purpose of raising beef for the butcher. Their value consists in their capacity to convert a large amount of highly nutritious food into a large amount of valuable beef, *and in the power they have of transmitting this quality to their offspring when crossed with ordinary cows*. It is in this last respect that pedigree is so important. But the former quality is due in a great degree to persistent high feeding for several generations. Were they submitted to ordinary food and treatment, especially when young, they would rapidly deteriorate. But put one of these splendid Shorthorn bulls to a carefully selected ordinary cow, and we get a grade Shorthorn that, with ordinary *good* feed and treatment, will prove highly profitable for the butcher.

The same is true of improved thorough-bred pigs. Their valuable qualities have been produced by persistent high feeding, and by selecting from their offspring those best adapted to high feeding. Pigs that grew slowly were rejected, while those which grew rapidly and matured early were reserved to breed from. In this way these qualities became established in the breed; and these qualities cannot be maintained without good care and good feeding.

In the case of pigs, we could well afford to give the necessary food to fatten thorough-bred pigs for the butcher. But we cannot afford to raise the young thorough-breds for this purpose. This would be true, even if we could buy thorough-bred sows and boars to breed from, at the price of ordinary pigs. The reason we cannot afford to raise highly refined, thorough-bred pigs for ordinary purposes, is, that if we feed them as they must be fed to maintain their qualities, they are apt to become too fat for breeding; and if we feed and treat them as ordinary slow-growing pigs are treated and fed, they lose the qualities which it is the object of the breeder to perpetuate. To raise highly improved, thorough-bred pigs, requires more care, skill, judgment, and experience than we can afford to bestow on animals designed to be sold in a few months to the butcher.

The object of raising improved thorough-bred pigs is simply to improve our common stock. They should be raised for this purpose, and for this purpose only. The farmer should buy a thorough-bred boar from some reliable breeder, and select the largest and best sows he has to cross him with. A thorough-bred boar at six weeks or two months old can usually be bought for $20 or $25. Such a boar in a neighborhood is capable of adding a thousand dollars a year to the *profits* of the farmers who use him.

CHAPTER VII.

GOOD PIGS NEED GOOD CARE.

We have said that an improved thorough-bred boar in a neighborhood is *capable* of greatly improving the qualities of the common stock, and adding largely to the profits of feeding pigs. But it is nevertheless a fact that such boars have been used by some farmers with little or no benefit.

There are several reasons for this: There are farmers in every neighborhood who half starve their breeding sows. Some of them do this deliberately, from a conviction that it improves their breeding and suckling qualities, just as some dairymen think a cow must be kept poor if she is to be a good milker. They mistake the cause for the effect. The cow is thin because she is a good milker, and not a good milker because she is thin. So a good sow gets very thin in suckling her pigs, but it is a great mistake to keep her thin, in order to make her a good breeder and suckler.

We have kept thorough-bred boars for some years, and have observed that those farmers who are liberal feeders speak highly of the cross, but those who believe in starving their sows, and letting the little pigs get their own living, assert that their pigs from a thorough-bred boar are no better than those from common boars.

The trouble is not in the thorough-bred boar, but in the sows. We use the improved thorough-bred boar in order to obtain pigs that will grow rapidly. But a pig cannot grow rapidly unless it has a liberal supply of food. It would be absurd to buy a superior mill, and then condemn it because it would not make choice family flour out of bran; and it is equally absurd to expect a pig, however perfect in form and fattening qualities, to make flesh and fat out of air and water.

A sow that has been starved all her life cannot produce vigorous, healthy pigs of good size, and with a tendency to grow rapidly and mature early. To put such a sow to an improved, thorough-bred boar, in hopes of getting good pigs, is as foolish as it is to hope to raise a large crop of choice wheat on wet, poor, neglected land, simply by purchasing choice seed. There is no such easy method of improving our stock. We must commence by adopting a more humane system of feeding, especially while the pigs are young. Then select the largest, thriftiest, and best-formed sows and put them to a good thorough-bred boar. Let the sow be regularly and liberally fed, without making her too fat. When with young she has a natural tendency to lay up fat, and it sometimes happens that a sow gets so fat that her pigs are small, and there is considerable danger of her lying on them. But there is far less danger from having a sow fat than is generally thought.

After she has pigged, feed the sow on warm slops, and other food favorable for the production of milk. Let the little ones be fed liberally, as soon as they commence to eat, and then the beneficial effect of using a good thorough-bred boar will be seen.

"But," it may be asked, "will not such liberal feeding produce good pigs without using a thorough-bred boar?" It will certainly produce better pigs than the starving system. But the effect of an improved thorough-bred boar in such a case is wonderful. We would rather pay $5 apiece for such pigs at two months old, than to accept as a gift, pigs from the same sow got by a common boar. At a year old we should expect the grades, in proportion to the food consumed, to bring at present prices, at least $10 a head more than the common stock.

We have a neighbor who is a good farmer, and who takes delight in feeding a good pen of pigs every fall and early winter. He "did not believe" in thorough-breds, and always spoke of my Essex, Berkshires, and Suffolks, as "nice *little* pigs." After watching the effect of a cross with good-sized common sows, he finally concluded to bring a young sow to one of our Essex boars. She was 16 months old, and certainly would not weigh over 120 lbs. It was then our turn to speak of *little* pigs. It so happened that we had a grade Essex sow the same age that accidentally took the boar at nine months old, and had a litter of nine pigs. She was very fat, and lay upon three of them. The remaining six were as handsome pigs as could be desired. These six pigs we sold at two months old, for feeding, for $35, and the sow, in a month after they were weaned, was killed, and dressed over 300 lbs., worth at the time 14 cts. per lb. or $42. Here then were two sows of the same age, one of which brought in $77, and the other at a liberal estimate was not worth $20. The difference was due simply to the use of a thorough-bred boar, and to liberal feeding. The one was half starved, under the mistaken impression that such treatment was best for breeding sows. The mother of the other was liberally fed, and her little ones were never starved.

During the summer, however, they had nothing but the wash and milk from the house, and the run of a good clover pasture. On this, the whole litter kept quite fat, and with the exception of this one sow, that proved to be with pig, were sold the first of October to the butcher, without having had any corn or grain of any kind for several months. The sow alluded to above, out of this litter, received the same treatment; but in a week or ten days after she pigged, we commenced to fatten her, and never did sucking pigs thrive better; and when they were weaned, the sow was actually fat, and in a month afterwards was *very* fat.

Now there is nothing remarkable in all this. We have had pigs do very much better, because better fed. But it certainly enabled us to silence the sneers of a prejudiced farmer against liberal feeding and thorough-bred pigs.

Another case deserves to be mentioned, showing the importance of liberal feeding in the case of well-bred pigs. One of our neighbors, a city man, who believed in good breeds and good feeding, had a common sow of good size and pretty fair form. He put her to a thorough-bred Prince Albert Suffolk boar, and had a litter of capital pigs. He afterwards put her to a thorough-bred Essex boar. But by this time, he got tired of farming, and at the sale, this sow was purchased by another neighbor who half starved her. She had a fair litter of pigs sometime in October. During the winter they had a little wash from the house and what they could pick up in a yard where cows received little or nothing but straw. The next summer they had the run of the roadside, with yokes around their necks to keep them out of mischief. A meaner and more utterly forlorn lot of pigs it has never been our lot to see. And this good man attributed his ill-luck to our thorough-bred boar!

In one sense he was right. The sow had been accustomed to liberal feeding, and the boar was descended from stock which, since the days of Lord Western and Fisher Hobbs, had been bred for the purpose of rapidly converting all the food they could eat into choice pork. No wonder that such a litter of pigs would not stand starvation as well as those more accustomed to it. Had the sow and the litter of pigs been liberally fed, they would have brought more money, with pork at 14 cts. per lb., than he received that year from his whole farm of 100 acres!

CHAPTER VIII.

THE ORIGIN AND IMPROVEMENT OF OUR DOMESTIC PIGS.

Nathusius has shown that all the known breeds of pigs may be divided in two great groups: one resembling, in all important respects and no doubt descended from, the common wild boar; so that this may be called the *Sus scrofa* group. The other group differs in several important and constant osteological characters; its wild, parent-form is unknown; the name given to it by Nathusius, according to the law of priority, is *Sus Indica*, of Pallas. This name must now be followed, though an unfortunate one, as the wild aboriginal does not inhabit India, and the best known domesticated breeds have been imported from Siam and China.*

Wild hogs still exist in various parts of Central and Northern Europe. The wild boar is described as having large tusks, a stronger snout, and a longer head than the domestic pig; smaller ears, pointed and upright; in color, when full grown, always black. He does not attain his full growth under four or five years, and will live for

---

* Darwin Animals and Plants Under Domestication, Vol. 1, page 85

Fig. 2.—WILD BOARS.

ORIGIN AND IMPROVEMENT OF OUR DOMESTIC PIGS. 43

twenty or thirty years. The sow breeds only once a year, and has seldom more than five or six at a litter; suckles

Fig. 3.—WILD BOAR.

them three or four months, and does not allow them to leave her until they are two or three years old, and able

Fig. 4.—ORIGINAL OLD ENGLISH PIG.

to defend themselves. Occasionally they grow to a great size, but usually they are not as large as the domestic pig.

The engravings in different parts of the book are, many of them, selected from different works, for the purpose of illustrating the changes which have been wrought in the hog by domestication, breeding, etc.

Great improvements have been effected by skillful breeders in the form of cattle and sheep, but we think these illustrations will show, that far greater improvement has been effected in the form of the pig than in any other animal. The picture of the "original old English pig" (fig. 4), shows a decided improvement in form over the Wild Boar (fig. 3). It has shorter legs, shorter head and

Fig. 5.—OLD IRISH PIG. *From Richardson.*

snout, heavier cheeks, a straighter and broader back, and larger hams. It will weigh more, in proportion to size, and afford more meat and less offal than the wild hog.

The engraving of the old Irish "Greyhound Hog" (fig. 5), shows an intermediate form between the wild and domestic animal. Richardson, from whose work the picture is taken, describes them as follows: "These are tall, long-legged, bony, heavy-eared, coarse haired animals; their throats furnished with pendulous wattles, and by no means possessing half so much the appearance of domestic swine as they do of the wild boar, the great original of the race. In Ireland, the old gaunt race of hogs has, for many years past, been gradually wearing away, and is now, perhaps, wholly confined to the western parts of the country, especially Galway. These swine are remarkably active, and will clear a five-barred gate as well as any hunter; on this account they should, if it is desirable to keep them, be kept in well-fenced inclosures."

The picture of the "original old English pig" shows that great improvement can be made merely by regular feeding and judicious selection; but it must be remembered that probably it took hundreds of generations to effect the change indicated in the engravings. That it could have been effected in a much shorter time, is undoubtedly true. But the fact remains that, centuries after the wild pigs had generally disappeared from the Island, the domestic pig derived from them was still a very coarse, slow maturing, and unprofitable animal.

The French and Germans, as compared with the English, have made but little improvement in the breeds of pigs, and many of the animals to be found upon the Continent are very much like the old English hog, bony, tall, gaunt, wiry-haired, and slow to fatten. On page 46 we give a portrait of a Craonnaire boar, which took a prize at a French agricultural show in 1856.

Fig. 6.—FRENCH PRIZE BOAR—CRAONNAIRE WHITE BREED.

CHAPTER IX.

IMPROVEMENT OF THE ENGLISH BREEDS OF PIGS.

The improvement in the breeds of pigs has kept pace with the improvement in general agriculture. High breeding is profitable when accompanied with high feeding and high farming; but a highly refined animal is not suited to a rude, primitive system of agriculture. The English breeds of pigs to-day, as compared with those of half a century ago, do not show greater improvement than is found in the general system of farming. There are still poor farmers in England, and there are also poor breeds of pigs; but it must be admitted that we can find in England the best specimens of high farming, and the best specimens of well-bred cattle, sheep, and pigs; and as good culture is rapidly becoming more general, there is an increasing demand for improved breeds, at high prices. There can be no doubt that the general improvement in agriculture, and the more general demand for improved breeds has greatly stimulated the efforts of the professional pig breeders; and it is doubtless true that several of the English breeds of pigs are to-day superior in form, early maturity, and fattening qualities, than any other breed in the world.

The early English breeders made great improvements, but being ahead of their times, they met with comparatively little demand for their improved pigs, and no adequate remuneration for their skill and labor.

It is not necessary to review the means employed by the breeders of the last century to improve the English breeds of pigs. Suffice it to say that it is generally admitted that much of this improvement is due to crossing the large English sows with the highly refined Chinese boars, and in selecting from the offspring such animals as possessed, in the greatest degree, the form and qualities desired. By continued selection, and "weeding out," the breed at length became established.

The Improved Berkshire is one of the earliest and best known of these Chinese-English breeds.

The old Berkshire hog had maintained a high reputation for centuries. It is described as "long and crooked

Fig. 7.—IMPORTED CHINESE SOW

snouted, the muzzle turning upwards; the ears large, heavy, and inclined to be pendulous; the body long and thick, but not deep; the legs short, the bone large, and the size very great." It was probably the best pig in England, and was wisely selected as the basis of those remarkable improvements which have rendered the modern Berkshire so justly celebrated.

It would be interesting to trace the different steps in this astonishing improvement, but, unfortunately, the nec-

IMPROVEMENT OF THE ENGLISH BREEDS OF PIGS. 49

essary information cannot be obtained. We give four engravings from Loudon's Encyclopedia of Agriculture, the

Fig. 8.—BERKSHIRE PIG.

first edition of which appeared in 1825, which will give some idea of the change that has been effected. Figure 8

Fig. 9.—HAMPSHIRE PIG.

is the Berkshire pig, as represented by Loudon, which is stated to represent "one of the best of its kind," and there can be little doubt that it was taken from what was considered a good specimen of the breed at the time the work was written. As compared with the figure of the old original English pig, and also with those of Hampshire, Herefordshire, and Suffolk, given by Loudon (figs. 9, 10, and 11), it is easy to trace the influence of the Chinese cross. Loudon speaks of the Berkshire, at that time, as a small breed, and it is undoubtedly true that the first effect of an improvement in the fattening qualities and early maturity of an animal is to reduce the size. On the whole, this picture of an improved Berkshire, forty-five or fifty years ago, does not give one a very favorable idea of the

Fig. 10.—HEREFORDSHIRE PIG.

breed at that time; yet it was then probably the best bred pig in England. Comparing this engraving with the one given by Youatt (fig. 12), in 1845, and with those given by Sydney in 1860 (figs. 20 and 21), we can form some idea of the remarkable effects of judicious breeding and high feeding. The engraving, figure 12, indicates the effect of a *cross* with the Chinese; the others show what can be done by persistent efforts in improving a breed of a mixed origin. It is highly probable that

IMPROVEMENT OF THE ENGLISH BREEDS OF PIGS. 51

boars of the improved Chinese-Berkshires, after the breed had become established, were employed to cross with the

Fig. 11.—SUFFOLK PIG.

large, old Berkshire sows, and that the effect of this less

Fig. 12.—BERKSHIRE SOW.

violent cross was more beneficial than the direct use of

the pure Chinese. Certain it is, that the pure Chinese pigs are now seldom, if ever, resorted to by English breeders. They find it more advantageous to resort to pure-bred boars of some of their own established breeds, although there is probably none of these breeds that have not, at one time or other, been crossed with the Chinese. It is a mistake, however, to speak of them, on this account, as "cross-bred" pigs, as is sometimes done. They have been bred pure long enough to become fully established.

The history of the Improved Essex Pig is of great interest, because better authenticated than that of any other breed.

The old Essex breed is described by Loudon as "up-eared, with long, sharp heads, roach-backed, carcasses flat, long, and generally high upon the leg, bone not large, color, white, or black and white, bare of hair, quick feeders, but great consumers, and of an unquiet disposition."

Lord Western, while traveling in Italy, saw some Neapolitan pigs, and came to the conclusion that they were just what he wanted to improve the breed of Essex pigs. He describes them, in a letter to Earl Spencer, as "a breed of very peculiar and valuable qualities, the flavor of the meat being excellent, and the disposition to fatten on the smallest quantity of food unrivaled." He procurred a pair of thorough-bred Neapolitans, and crossed them with Essex sows, and probably with black Sussex and Berkshires. He obliterated the white from the old Essex, and obtained a breed of these cross-bred pigs that could scarcely be distinguished from the pure-bred Neapolitans.

These Neapolitan-Essex had great success at agricultural fairs, but as Lord Western continued to breed from his own stock, selecting the most highly refined males and females, they "gradually lost size, muscle, and constitution, and consequently fecundity; and at the time of his death, in 1844, while the whole district had benefited from the cross, the Western breed had become more ornamental than useful."

In other words, while this highly refined breed was of great value to cross with the large, vigorous sows in the neighborhood, they were not profitable to raise pure. This is the case with all highly refined, thorough-bred pigs. They are not as profitable for the mere production of pork as the pigs from the common sow and a thorough-bred boar. It is as true to-day as it was then, that any highly refined throrough-bred pigs are "more ornamental than useful," unless farmers know how to use them; then they are of great value. In the meantime, a tenant farmer of Lord Western, the late Fisher Hobbs, of Boxted Lodge, had availed himself of the opportunity to use the thorough-bred Neapolitan-Essex boars belonging to Lord Western, and crossed them with the large, strong, hardy, black, and rather rough and coarse Essex sows, and in process of time he established the breed, since become so famous—the Improved Essex.

The difference between the two breeds is shown by the engraving of one of Lord Western's Neapolitan-Essex boars (fig. 23), drawn for the first edition of Youatt on the Pig, and that of "Emperor" (fig. 22), an eight-year-old Improved Essex working boar, taken in 1860 for Sidney's last edition of Youatt on the Pig.

Sidney, in his last edition of Youatt, says: "The Improved Essex probably date their national reputation from the second show of the Royal Agricultural Society, held at Cambridge, in 1840, when a boar and sow, both bred by Mr. Hobbs, each obtained first prizes in their respective classes.

"Early maturity, and an excellent quality of flesh, are among the merits of the improved Essex. They produce the best 'jointers' for the London market. With age they attain considerable weight, and often make 500 lbs.

at twenty-four months old. 'Emperor' (fig. 22) is 2 ft. 8¼ in. high at the shoulder, and 6 ft. 1 in. long. Boars bred at Boxted have been known to reach 36 in. in height.

"The defect of the improved Essex is a certain delicacy, probably arising from their southern descent, and an excessive aptitude to fatten, which, unless carefully counteracted by exercise and diet, often diminishes the fertility of the sows, and causes difficulty in rearing the young. As before observed, they are invaluable as a cross, being sure to give quality and early maturity to any breed, and especially valuable when applied to a black breed where porkers are required. For this purpose they have been extensively and successfully used in all the black pig districts of this country [Great Britain], where, as well as in France and Germany, and in the United States, they have superseded the use of the imported Neapolitan and Chinese. Many attempts, on a limited scale, to perpetuate the breed pure, have been unsatisfactory, because it is too pure to stand in-and-in breeding. They require much care when young. 'In the sows the paternal fattening properties are apt to overbalance the milking qualities, and make them bad nurses.'

"The Berkshire breed have benefited much from the improved Essex cross. The best Devonshire pigs have a large infusion of the same strain. The improved Dorsets, the most successful black pigs ever shown at the Smithfield Club shows, have borrowed their heads at least from the Essex breed. The improved Oxfords are the result of a judicious blending of pure Neapolitan, Berkshire, and improved Essex blood; and throughout the midland and western counties, the results of Lord Western's Italian tour are to be found in every parish where a black pig is patronized.

"The history of this breed affords a good illustration of the advantages of the system under which landlords, stimulated by patriotism or competition, or mere love of things agricultural, breed and experiment with great zeal, varied success, and little or no profit, until they reach the point where the tenant farmer, with sufficient capital, equal zeal, and a clear eye to the £. *s. d.*, takes up the work, breeds, and works the problem out with a degree of practical knowledge, personal attention, and enthusiasm, which few, except farmers breeding for a profit, can contrive to combine, and persevere to bestow for a long series of years.

"Foreign governments endeavor, with very limited success, to produce the effect of our aristocratic breeding enthusiasts by government studs. But an official, however gilded, titled, or crossed, has never the influence of a peer or squire; and besides his name, the raw materials—the working bees, the great tenant farmers—are wanting on the continent.

"The improved Essex are ranked amongst the small breeds, and there they are most profitable; but exceptional specimens have been exhibited at agricultural shows in the classes for large breeds, as, for instance, at Chelmsford, in 1856.

"There is probably no black pig which combines more good qualities, as either porker or bacon hog, than the produce of an improved Essex boar and an improved Berkshire sow."

The facts here narrated are of great importance as illustrating the principles of breeding which we have endeavored to lay down in the first chapters of this work. The old Essex pigs were *great eaters*. All the authorities mention this fact as one of the *objections* to the breed. The Lord Western Essex were highly refined pigs, of good form, little offal, maturing very early, and fattening with great rapidity, but destitute of size and vigor. Crossed with selected sows of the old, hardy, vigorous race, the offspring possessed the form, early maturity, and fattening qualities of the improved breed, united with the constitution of the common stock. They had the stomach of the mother, and the refinement of the sire. No wonder that they "have an excessive aptitude to fatten." What else can they do with the large amount of food they are capable of eating and digesting except to convert it into flesh and fat? There is a minimum of offal in this breed, and they are exceedingly quiet. There is little demand on the large quantity of food they can eat, and nearly the whole of it must be converted into flesh and fat; and we have endeavored to show the immense advantage of having an animal that will consume a considerable excess of food over and above that required to sustain the vital functions. In this view of the matter it is easy to see why the Improved Essex proved such a useful breed in the hands of intelligent farmers.

Many other similar instances of the improvement of English breeds might be given, but it is not necessary to do so. The principle which underlies them all is the same. A large, vigorous, healthy sow, crossed with a highly refined, thorough-bred boar, and the offspring carefully bred until the desired qualities become established in the new or improved breed.

CHAPTER X.

THE MODERN BREEDS OF ENGLISH PIGS.

English writers on swine, twenty years ago, describe a dozen or more breeds of pigs, then kept in England, and nearly as many more in Scotland and Ireland. Youatt and Richardson, both of whose works on the pig were reprinted in this country, give a full account of these old breeds. Many of these breeds have been, at one time or another, introduced into the United States and Canada;

but comparatively few of them have been kept pure, either here or in England. The common stock of pigs in America is made up of these old breeds. Occasionally we see a pig that has some distinct characteristics recognizable as belonging to some known breed, but, as a general rule, it is impossible to trace the slightest resemblance to any distinct breed, either of the past or the present.

The same is true, to a considerable extent, in England. The common stock of pigs is of such a mixed character, that it can be traced to no particular breed. Many of the old breeds have become extinct. We have so-called "Cheshire" pigs in America, but there is no such breed raised or known in Cheshire, and has not been for twenty years or more.

Culley, in his work entitled "Observations on Livestock," published in 1807, gives a well authenticated account of a Cheshire pig which measured, from the nose to the end of the tail 9 feet, 8 inches, and in height, 4 feet, 5½ inches. When alive, it weighed 1,410 lbs., and when dressed, 1,215 lbs. The age is not given. It was probably as fat as it could be made, and yet it only dressed 80½ per cent of its live weight.

This breed, if we may call it a breed, was evidently very large and coarse. It is described as "remarkably long, standing very high, on long, bony legs; head large, ears long and hanging; back much curved, and narrow; sides flat and deep; color white, blue and white, black and white." This breed has become extinct.

The old Yorkshire or Lincolnshire breed is described in Morton's Cyclopedia as "one of the largest breeds in the kingdom, and probably one of the worst; extremely long-legged, and weak loined; very long from head to tail; color chiefly white, with long, coarse, curly hair; tolerable feeders, but yielding a coarse, flabby flesh, of inferior marketable quality."

It is from this race of pigs that the modern Yorkshire,

Fig. 18.—"SIR ROGER DE COVERLY." YORKSHIRE LARGE BREED.

now perhaps the most popular breed in England, has been derived. This breed is divided into three classes: the Large, Medium, and Small.

THE LARGE YORKSHIRES.—(Figs. 13, 14 and 15.)

We have shown what the old Lincolnshire and Yorkshire pig was before any especial efforts had been made to improve it. In 1854, Mr. A. Clarke, of Long Sutton, Lincolnshire, the author of a valuable treatise on the breeding and management of pigs in Morton's Cyclopedia of Agriculture, writes: "In the adjoining county of Yorkshire the breeders have outdone the Lincolnshire breeders in point of size, but not in any other respect. The specimens lately exhibited at our meetings, of the large Yorkshire breed, by Messrs. Abbot, Taylor, Tuley, and others, have attained a size too large for any useful purpose, and would exceed in weight that of a moderately grown Scotch ox. The present taste of the public is decidedly set against such an overgrown sort; at present, however, they make large prices." We believe there is now no breed known as the Lincolnshire. It has been merged in the Yorkshire.

Of the old, unimproved large Yorkshire, Sidney says: "It was a long time coming to full size, and could be fed up to 800 lbs., but whether with any profit, is doubtful. It was and is still very hardy, and a very prolific breeder. Attempts have been made to improve it by crossing with the Berkshire, Essex, Neapolitan, and other black breeds, which produced a black and white race. Those from the Berkshire are a hardy, useful sort, but fatten slowly; the other crosses have little or no hair, are too delicate for the North, and are fast wearing out.

"The first step taken in the right direction for improving the old Yorkshire seems to have been the introduction of the White Leicesters. These were a large sort, with

Fig. 14.—"PARIAN DUCHESS." YORKSHIRE LARGE BREED.

smaller heads than the old York, erect ears, finer in the hair, and lighter in the bone.

"The improvement in the York large breed commenced early in the century, when the White Leicesters were introduced. The general run of pigs in the grain-growing districts of Yorkshire shows that they partake more or less of this cross. The old sort is seldom seen except in the northern part of the county."

A Yorkshire correspondent of Mr. Sidney, writing in 1860, says "The Leicester cross has been still further improved by putting the largest and best sows of the Leicester cross to boars of the small white breed from Castle Howard* and Bransby†, breeding from the progeny, and selecting the largest and best of the young sows and the best formed boars for that purpose, taking care that they were not too nearly related. By this means the size and constitution of the large breed, with the symmetry and tendency to fatten of the small breed, have been, in a great degree, transmitted to the offspring. If a sow shows too much of the old sort, she is put to a boar of the small breed for her first litter."

Such seems to have been the origin of the present breed of Large Yorkshires.

"These improved Large Yorkshires," (says Sidney, in 1860,) "are principally bred in the valley of the Aire, in the neighborhood of Leeds, Keighley, and Skipton. They are in great request as breeding stores, and purchased for that purpose for every part of the United Kingdom, as well as for France, Germany, and the United States, at great prices."

These pigs "can be fed to 60 stone, of 14 lbs., dead weight, or 840 lbs. The Prize Boar at the Royal Agricultural Fair at Chester weighed, alive, 1,232 lbs. The

---

* The Earl of Carlisle.

† Mr. Wyley, of Bransby, introduced a small breed of White Leicesters, now called Yorkshires.

Fig. 15.—"GOLDEN DAYS." YORKSHIRE LARGE BREED.

Prize Sow at the Royal Fair at Warwick, 1,204 lbs. At Northallerton, in 1859, the finest lot of large sows ever seen in one place were collected together. There were at least a dozen, each of whose live weight would not be much less than half a ton (1,120 lbs). The Royal Agricultural Prize-winner at Norwich was only just good enough to get second honors."

Mr. Wainman, the owner of Carhead Farm, has been one of the most successful breeders of the Large Yorkshires, having won more than two hundred prizes, and sold, in the language of one of his Yorkshire admirers, the produce of one sow "for as much as would build a church." Mr. Fisher, who is bailiff at Carhead Farm, gives the weight of two of these pigs. One, killed at less than 7 months, dressed 255 lbs., and one at 12 months old, 489 lbs.

THE SMALL YORKSHIRES.

Mr. Mangles, "one of the first pig-breeders and feeders in Yorkshire," gives the following description of the Small Yorks: "The small Yorkshire is peculiar to Yorkshire, and different from any other breed I have seen. It has a short head, small, erect ears, broad back, deep chest, and short legs, with fine bone. It is always ready to fatten, and turn to account either in the way of roasters, small porkers, small bacon, or medium. Three or four of the small breed might be fed well, and kept fresh and symmetrical on the food which would barely keep one lean and gaunt large Yorkshire."

THE SMALL CUMBERLAND.

"The Cumberland small breed," says Mr. Sidney, "are described by Mr. Brown, of Aspatria, who is one of the most noted founders of the modern breed, from whom Lord Ducie purchased some of his most celebrated animals, as not small in reality, but a medium size, short in the

Fig. 16.—CUMBERLAND YORK BOAR. SMALL BREED.

legs, back broad, straight, and evenly fleshed; ribs well developed, rumps and twists good; hams well down, and low; breast and neck full, and well formed; no creases in the neck; ears clean, fine, of a moderate size, and standing a little forward; nose short; body evenly covered with short, fine hair."

At the Birmingham show, in 1850, Mr. Brown won all first prizes in small breeds for the "best boar," "best sow," and "best pen of pigs," with his Cumberland breed; and sold a boar and sow under six months old for forty-three guineas to Earl Ducie. At the sale, on the death of the Earl, the sow "Miss Brown" was sold to the Rev. F. Thursby for sixty-five guineas. "She paid me," he writes to Mr. Sidney, "very well, having sold her produce for £300, and have now (February, 1860,) four breeding sows from her."

THE YORK-CUMBERLAND BREED (fig. 16).

Mr. Sidney classes the Small Yorkshire and Cumberland together, "because, although originally, they somewhat differed in size,—the Cumberland being the larger—they are being continually intermixed, with mutual advantage; and pigs of exactly the same form, the result of crosses, are constantly exhibited under the names of Yorkshire or Cumberland, according to the fancy of the exhibitor."

Mr. Mangles writes—"The Small Cumberland is a great deal larger than the Small Yorkshire. By judiciously crossing the two, I have obtained a breed combining size, aptitude to fatten, and early maturity. From the Cumberland I got size, and from the Yorkshire quality and symmetry. I have tried a great many breeds of pigs, and, keeping the pounds, shillings and pence in view, have found no breed equal to the Yorkshire and Cumberland cross."

A Warwickshire correspondent of Mr. Sidney writes:

"No animal of the pig species carries so great a proportion of flesh to the quantity of bone, or flesh of as fine a quality, as the small Yorkshire, or can be raised at so small a cost per pound. With common store food they can always be kept in condition—with common care, and slight addition to food, they are ready to be killed, for porklets, at any age; and if required for bacon, take one

Fig. 17.—PRIZE YORK-CUMBERLAND PIG. SMALL BREED.
From Farmers' Magazine.

farrow of pigs from a yelt.* You ought to have from seven to ten pigs the first time. I have four sisters, yelts, that have brought me thirty-eight pigs this last January. They are as pure as 'Eclipse,' being descended from the

---

* A "yelt" is a young sow before she has had pigs. The idea here is, when it is desired to obtain bacon from the small breeds, to take one litter of pigs from a young sow, and then fatten her. Ordinarily, it will not pay to keep these pigs long enough to make large pork: but if a litter of pigs can be obtained in the meantime, it is then very profitable. But if we should continue to breed from pigs of the first litter, the size would soon become too small.

stock of Earl Ducie and Mr. Wyley, of Bransby, near York, and are of good size. I killed a sow this winter that weighed 26 score—520 lbs.

"The ordinary weight is from 14 to 17 score—280 lbs. to 340 lbs. In some cases, where very thick bacon is required, they may be profitably got to 30 score—600 lbs. The Small Yorkshire owes its present superiority to choice selections, and judicious crossing of different families of the same breed; by this means size is maintained with character."

These "Small Yorkshires" which this gentleman calls as "pure as Eclipse," are descended from the stock of Earl Ducie and Mr. Wyley; but, as has been already shown, Earl Ducie purchased Cumberland pigs from Mr. Brown, and Mr. Wyley's original stock were White Leicesters.

Mr. Sidney says: "The wide extension of this Cumberland and York blood is to be traced wherever the Royal Agricultural Society's prizes for white pigs are won.

"Thus:—Mr. H. Scott Hayward, of Folkington, a prize-winner at Chelmsford, in 1856, in small breeds, with a white sow, states that he has used boars from the following breeders:—The late Earl of Carlisle, Castle Howard; the late Earl of Ducie; the Earl of Radnor, Coleshill; and at present (1860) one from the Prince Consort's stock.

"The card of Mr. Brown's boar 'Liberator' contains the following pedigrees, and shows a distinct connection between Cumberland and Yorkshire, and all the most celebrated white breeds in the south :—

"'Liberator' was bred by Earl Ducie, got by 'Gloucester,' dam 'Beauty' by Lord Radnor's boar, gr.-d. 'Julia Bennett' by Lord Galloway's boar, etc.

"'Gloucester' was bred by the Earl of Ducie, got by 'General,' dam 'Hannah' by the 'Yorkshireman;' gr.-d. bred by the Earl of Carlisle, and purchased by Lord Ducie at the Castle Howard sale.

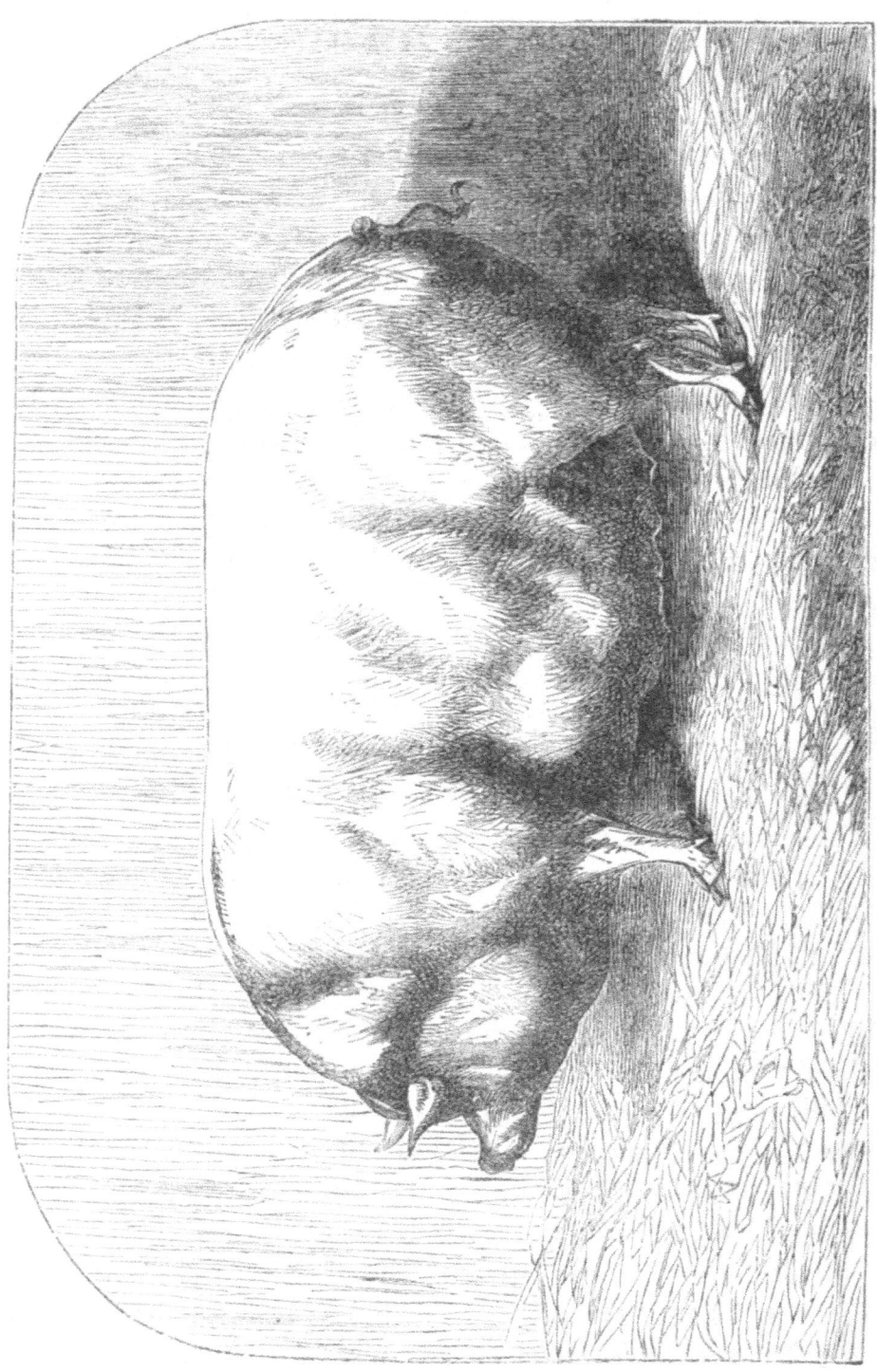

Fig. 18.—"MISS EMILY," YORKSHIRE MIDDLE BREED

"'General,' bred by Mr. Wyley, sold to Mr. Mackintosh, of London, and hired by H.R.H. Prince Albert, the Earl of Ducie, and Lord Wenlock, and was the sire of two pens of pigs, the property of H.R.H. Prince Albert, that obtained the first prize at a Smithfield Christmas Show.

"It may, therefore, safely be assumed that all the best white pigs of modern times have been bred from Yorkshire or Cumberland and white Leicesters, or both; and many breeds, such as Middlesex, Coleshill, etc., may be dismissed as mere variations of the white small Yorkshire.

"Mr. G. Mangles, of Givendale, near Ripon, Mr. Brown writes me, was one of the first to cultivate the cross of the York-Cumberlands."

THE MIDDLE OR MEDIUM YORKSHIRE BREED (Fig. 18).

"The Yorkshire medium or middle breed," says Mr. Sidney, "is a modern invention of Yorkshire pig-breeders, and perhaps the most useful and the most popular of the white breeds, as it unites, in a striking degree, the good qualities of the large and the small. It has been produced by a cross of the large and the small York, and the Cumberland, which is larger than the small York. Like the large whites, they often have a few pale-blue spots on the skin, the hair on these spots being white. All white breeds have these spots more or less, and they often increase in number as the animal grows older.

"It was not until 1851 that the merits of this breed were publicly recognized at a meeting of the 'Keighley Agricultural Society,' when, the judges having called the attention of the stewards to the fact that several superior sows, which were evidently closely allied to the small breed, had been exhibited in the large-breed class, the aspiring intruders were, by official authority, withdrawn.

"They included the since celebrated 'Sontag,' 'Jenny

Lind,' 'Kick-up-a-dust,' and some other distinguished grunters, forming altogether such an imposing *troupe*, that the authorities gave them a performance (*i. e.*, a class) to themselves, with a benefit in the shape of first and second prizes, and called them the '*middle breed.*'

"This example was generally adopted throughout Yorkshire, and at local shows they are the strongest and best-filled of all the classes.

"The principal prize-takers amongst the boars in this breed have been 'Paris,' 'Nonpareil,' 'Lord Raglan,' 'Sir Colin,' and 'Wonder;' and amongst the sows, 'Zenobia,' 'Lady Airdale,' who held her own during two seasons, in one of which she took ten prizes, 'Craven,' 'Lady Kate,' 'Queen Anne,' and '*Miss Emily*' (see portrait), who has never found her marrow, having taken nine first prizes in succession, including the champion cup at Caldervale show in 1859, for the best pig in all classes. This competition brought all Yorkshire, several Warwick, Royal Highland Society, Dublin and Irish Royal, as well as Cheshire and Lancashire champions, to the Cloth Hall, Halifax. Amongst the rest, 'CARSWELL,' the second winner in the *large* boar class at Warwick, entered in the middle class, and carried off the first prize in that class; but in the trial for the championship, he was beaten like the rest, and the plate, with the 'white rosette of York,' went to 'MISS EMILY,' whose girth, taken behind the shoulder, was at this time eighty-five inches. She fully qualified for all the prizes she had taken as a *breeding sow*, by producing at Carhead the following October a fine litter of pigs.

"The middle Yorkshire breed are about the same size as the Berkshire breed, but have smaller heads, and are much lighter in the bone. They are better breeders than the small whites, but not so good as the large whites; in fact, they occupy a position in every respect between these two breeds."

THE MODERN ENGLISH BREEDS OF PIGS. 71

Fig. 19.—WHITE LEICESTER BOAR AND SOW. SMALL BREED.

WHITE LEICESTERS (Fig. 19).

We can ascertain nothing satisfactory in regard to the origin of this breed of pigs. This is the more to be regretted as the fact that they were "the great improvers of the gigantic Yorks," invests them with more than ordinary interest.

Mr. J. W. Williams, of Somersetshire, is the principal breeder of White Leicesters. He first exhibited in 1852, and has taken the Smithfield Club gold medal, two gold medals at the Paris Exposition in 1855, and numerous other prizes. The portrait of the Paris Prize Leicesters is given on page 71 (fig. 19). Mr. Williams states that his fat pigs of this breed generally average the following weights:

```
5 to  6 months,  7 to  9 score lbs..............140 to 180 lbs.
    8       "   10 to 12   "    "  ..............200 to 240  "
   10       "   12 to 15   "    "  ..............240 to 300  "
12 to 18    "   15 to 18   "    "  ..............300 to 360  "
```

The pen of three pigs of this breed which received the Smithfield Club gold medal in 1854 weighed, sinking offal, at 18 weeks old, 180 lbs. each.

SUFFOLK AND OTHER WHITE BREEDS.

Mr Sidney says: "Yorkshire stands in the first rank as a pig-breeding county, possessing the largest white breed in England, as well as an excellent medium and small breed, all white, the last of which, transplanted into the south, has figured and won prizes under the names of divers noblemen and gentlemen, and more than one county. The Yorkshire are closely allied with the Cumberland breeds, and have been so much intermixed that, with the exception of the very largest breeds, it is difficult to tell where the Cumberland begins, and where the Yorkshire ends. It will be enough to say, for the present, that the modern Manchester boar, the improved Suffolk, the improved Middlesex, the Coleshill, and the Prince Alberts or Windsors, were all founded on Yorkshire-Cumberland stock, and some of them are merely pure Yorkshires transplanted, and re-christened."

Speaking of the pigs kept in the dairy district of Cheshire, he says: "White pigs have not found favor with the dairymen of Cheshire, and the white ones most used are 'Manchester boars,' *another name* for the Yorkshire-Cumberland breed. 'Mr. Youatt,' he says, in another place, 'and all the authors who have followed him, down to the latest work published on the subject, occupy space in describing various county pigs which have long ceased to possess, if ever they possessed, any merit worthy of the attention of the breeder. Thus the Norfolk, the Suffolk, the Bedford, the Rudywick, the Cheshire, the Gloucester breeds, have each a separate notice, not one of which, except the Suffolk, is worthy of cultivation, and the Suffolk *is only another name for a small Yorkshire pig.*"

BLACK BREEDS.

If all the modern white breeds in England, of any special value to the breeder, are Yorkshires, or York-Cumberland and Leicesters, it is equally true that there are but two breeds of black pigs that deserve any special attention—the Essex and Berkshire.

"Black pigs and their crosses," says Mr. Sidney, "occupy almost exclusively the counties of Berks, Hants, Wilts, Dorset, Devon, and Somerset. Sussex has a black county breed, and in Essex a black-and-white pig has become all black. In the Western counties, the prejudice against a white pig is nearly as strong as against a black one in Yorkshire. In Devonshire, white pigs are supposed to be more subject to blistering from the sun when pasturing in the fields.

"For breeding purposes, the black breeds may be divided into two—the improved Berkshire and the improved Essex, because there is no dark breed that has special characteristics so well worth cultivation as these two, and there is no black pig that may not be advantageously crossed by boars of one or both of these breeds. Hampshire has an ancient, coarse, and useful breed of black pigs. They are inferior to Berkshire, and not in the same refined class as Essex, therefore not worth taking from their native county"

BERKSHIRE.

" Among the black breeds," says Mr. Sidney, " by universal consent, the improved Berkshire hog stands at the head of the list, either to breed pure, or to cross with inferior breeds. The Berkshire was originally a large breed (it has very recently carried off prizes in the large classes at Royal Agricultural and other shows) of a black-and-white and sandy-spotted color, as represented in the portrait given by Mr. Youatt (fig. 12), in this respect distinctly differing from its neighbor, the old black Hampshire hog, rather coarse, but of general form very superior to the old white and black-and-white farm hog of the northern counties.

" The late Lord Barrington (who died in 1829) did a great deal towards improving the Berkshire breed, and the improved Berkshires are almost all traced back to his herd. They are now considered by Berkshire farmers to be divided into middle (not a large breed) and a small breed. If first-class, they should be well covered with long black silky hair, so soft that the problem of 'making a silk purse out of a sow's ear' might be solved with a prize Berkshire. The white should be confined to '*four white feet, a white spot between the eyes, and a few white hairs behind each shoulder.*'

THE MODERN ENGLISH BREEDS OF PIGS.

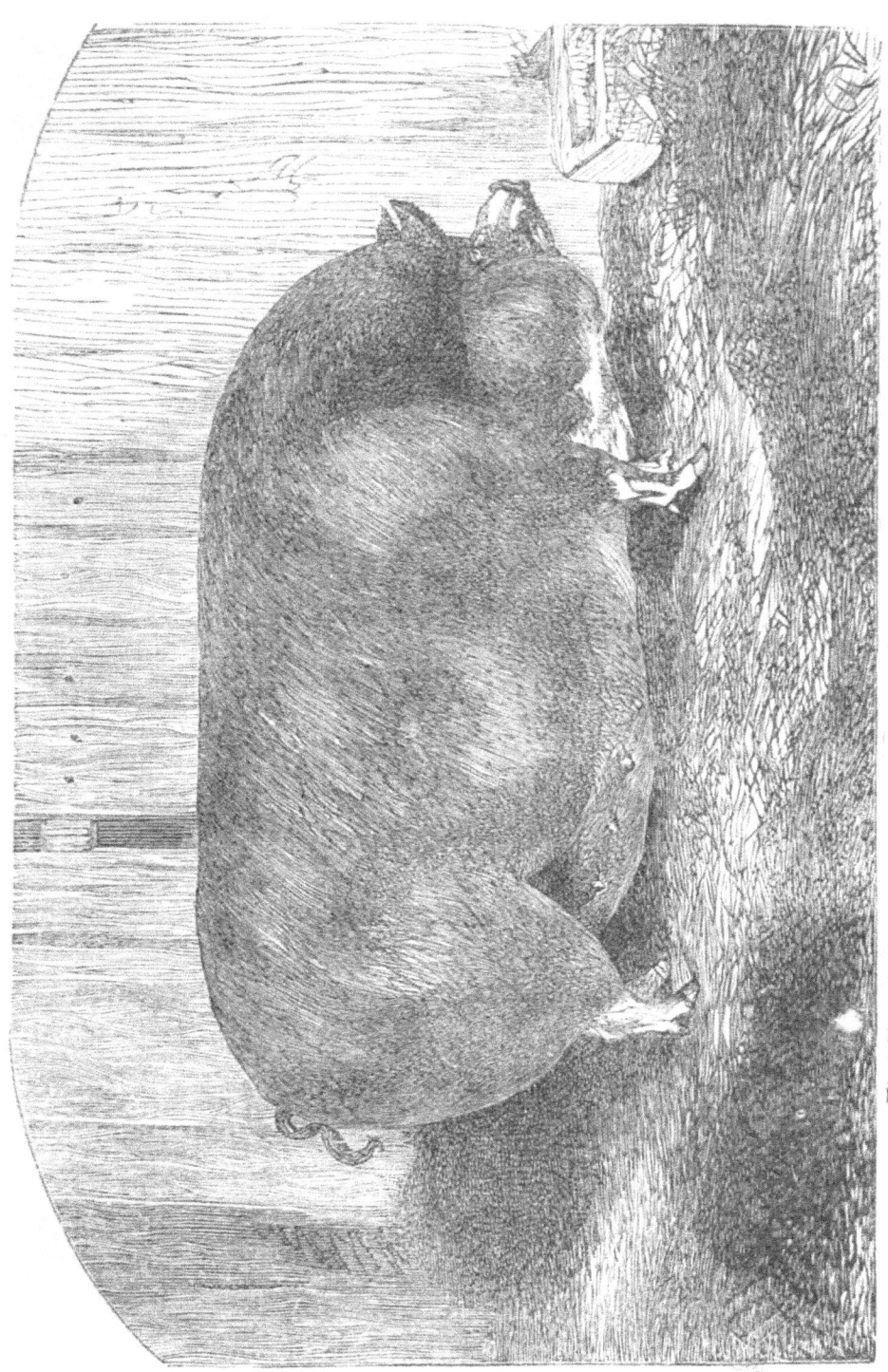

Fig. 20.—SMITHFIELD CLUB PRIZE FAT SOW. IMPROVED BERKSHIRE.

"At Mr. Sadler's, Bentham, near Cricklade, one of the most successful improvers of Berkshires, and eminent as a manufacturer of North Wiltshire cheese, the committee of the Ayrshire Agricultural Association saw 'three hundred, every one of which was marked in this manner.'

"Mr. Sadler obtained his original stock from the late Lord Barrington's herd. At Baker Street, he once won the prize for the best fat pig in the yard with a sow nearly four years old, (a portrait of which is given in fig. 20,) which had been the mother of a numerous progeny. She was 6 ft. 4 in. in length, 7 ft. 6 in. in girth, and weighed 42 score, 16 lbs., or 856 lbs.,—more than many fat heifers. But it seems to be the general opinion of feeders that Berkshires pay best at moderate weights.

"To develop the full size, they must not be allowed to breed until twelve months old at least. Mr. Sadler considers the improved Berks superior to any other (black?) breed, for size, quality, hardiness of constitution prolificness, early maturity, and aptitude to fatten.

"My friend Mr. Thomas Owen, of Clapton, Hungerford, who has had, in his forty years' experience as a Berkshire farmer, 'some thousand through his hands dead,' writes me:

"'I remember the Berks pig a much larger and coarser animal than now; at present they are a medium, not a large breed. They have been improved by judicious selection and distant crosses with the Neapolitan, which have added to their fattening qualities. They are much esteemed by butchers for evenness of flesh (that is, more lean to the proportion of fat) than any other breed,—and this is a good recommendation.'

"The late Rev. T. C. James, who was a successful exhibitor of pigs at Chelmsford, and one of the judges of pigs at the Royal Agricultural Society's show at Warwick, in 1859, wrote: 'The improved Berkshire is a good big animal, well calculated to produce a profitable flitch. A

THE MODERN ENGLISH BREEDS OF PIGS. 77

Fig. 21.—IMPROVED BERKSHIRE BOAR. MIDDLE BREED.

good little pig is very well, but a good big pig is better, if with aptitude to fatten: two exhibited at Chelmsford, in 1856, (of Sadler's breed) weighed, each, twelve score at seven months old, and with that weight, were of such good constitution, that they were well upon their legs. They had walking exercise in an orchard every day while fattening.'

"One of the most extensive farmers in West Norfolk writes: 'Dissatisfied with the Norfolk pigs, I flew to Mr. Sadler, of Bentham, Wilts, gave him 20 guineas for three sows and a boar. I sold over one hundred in the first eighteen months for £2 each when ten weeks old, and the only complaint I have is, that they do not breed so many as the old Norfolks; but I say eight or nine good ones are better than ten or eleven ordinary ones. They are good graziers, and our butchers are very fond of them. There is plenty of *lean meat* with the fat, which is not the case with the fancy pigs. The cross between the Berks boar and Norfolk sow (white), like all cross breeds, is most profitable to the feeder, but we must have pure breeds first.'

"This Norfolk opinion," says Mr. S., "is confirmed by all my correspondence. The Berkshire pig is in favor in every dairy district, either pure or as a cross, but chiefly as a cross; he does not fatten so quickly as some other breeds, but his constitution and bacon quality are famous.

"The average weight of a bacon improved Berkshire hog, fit to kill, will be about 400 lbs. The ham-curers who purchase from these farms, prefer the small breed of Berkshires, of from nine to fourteen score.

"The improved Berkshire boar was used to give size and constitution, many years ago, to the Essex; and the most eminent breeder of Essex has informed me that on one occasion, in a litter of Essex pigs, two little pictures of the Berkshire boar, their remote ancestors by at least twenty-eight years, appeared. It seems to be generally

THE MODERN ENGLISH BREEDS OF PIGS.

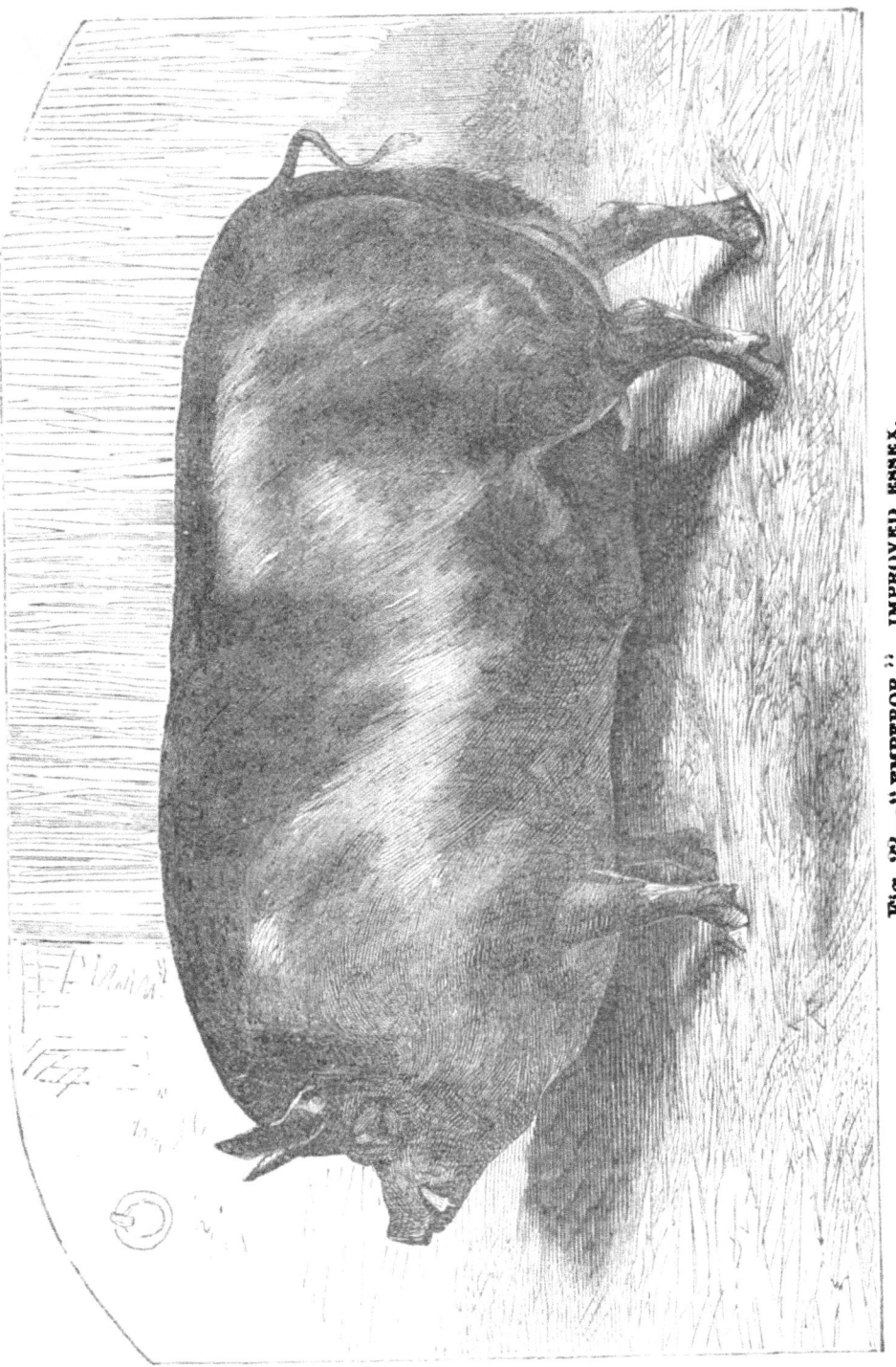

Fig. 22.—"EMPEROR." IMPROVED ESSEX.

agreed that the Berks breed is best adapted for hams and bacon, and not for small fresh pork. As I have already mentioned, the Berks boar has been used to cross the large breed in Yorkshire, but without permanently satisfactory results in establishing a breed; for a first cross with almost any breed, it is sure to produce a well-sized useful animal. In reply to questions addressed through the landlord of the Arley Hall estate, in Cheshire, to his principal tenants, it seems that the dairy farmer of that county finds it profitable to cross the dark or spotted sows which they have in the county, and also those they purchase largely from Shropshire and Wales, with a Berkshire boar. The produce is all, or nearly all, made into, and sold for making bacon. On the other hand, in Kent, Mr. Betts, of Preston Hall, buys Berkshire sows and crosses them with a white Windsor boar, 'the produce being invariably white."

IMPROVED ESSEX (Fig. 22).

We have already given some account of this celebrated breed, but the American farmer will be glad to read what Mr. Sidney writes in regard to it. He says: "The improved Essex is one of the best pigs of the small black breeds, well calculated for producing pork and hams of the finest quality for fashionable markets; but its greatest value is as a cross for giving quality and maturity to black pigs of a coarser, hardier kind. It occupies, with respect to the black breeds, the same position that the small Cumberland-Yorks do as to white breeds—that is to say, an improved Essex boar is sure to improve the produce of any large dark sow.

"The original Essex pig was a party-colored animal, black, with white shoulders, nose, and legs—in fact, a sort of 'sheeted' pig, large, upright, and coarse in bone.

"The first improvement was made by the late Lord Western, when Mr. Western, an Essex squire, who divided his life pretty equally between the cultivation of live-stock and the passionate support of the politics of his friend, Charles James Fox. While traveling in Italy (making the grand tour), he observed, admired, and secured a male and female of the breed called Neapolitan, 'found in its greatest purity (according to a letter addressed by Lord

Fig. 23.—LORD WESTERN ESSEX.

Western to Earl Spencer in the *Farmers' Magazine*, January, 1839) in the beautiful peninsula, or rather tongue of land, between the Bay of Naples and the Bay of Salerno. . . A breed of very peculiar and valuable qualities, the flavor of the meat being excellent, and the disposition to fatten on the smallest quantity of food unrivaled.'

"From this pair Mr. Western (afterwards Lord Western) bred in-and-in, until the breed was in danger of becoming extinct—a sure result of in-and-in breeding. He then turned to Essex, and, there is reason to believe, to black

Sussex and Berkshire sows; and obliterating the white of the old Essex, produced a class of animals of which he says, in the letter already quoted: 'I have so completely engrafted this stock upon British breeds, that I think my herd can scarcely be distinguished from the pure blood" (of Neapolitans). (See figure 23.)

"The Western Essex pigs had great success at agricultural shows. The old Essex, with its 'roach back, long legs, sharp head, and restless disposition,' was capable of being made very fat, but then it required time and an unlimited supply of food. The advantage of a cross with the Italian was obvious, and the fact that the new breed was in the hands of a popular county squire was no small help in extinguishing the native and unprofitable particolored race.

"But as Lord Western bred exclusively from his own stock—having attained what he considered perfection—always selecting the neatest and most perfect males and females, his breed gradually lost size, muscle, and constitution, and consequently fecundity; and at the time of his death, in 1844, while whole districts had benefited from the cross, the Western herd had become more ornamental than useful.

"But, in the meantime, the well-known Mr. Fisher Hobbs, of Boxted Lodge, then a young tenant farmer at Mark's Hall, on the Western estate, had taken up, among other farm live-stock, the Essex pig, and made use of the privilege he enjoyed of using Lord Western's male animals to establish a breed on strong, hardy black Essex sows, even if somewhat rough and coarse, crossed with the Neapolitan-Essex boars. On the carefully selected produce of these, divided and kept as pure separate families, he established the breed that he first exhibited, and has since become famous as the '*Improved Essex*,' a title which Lord Western himself adopted when his tenant and pupil had successfully competed with him. On Lord

THE MODERN ENGLISH BREEDS OF PIGS.

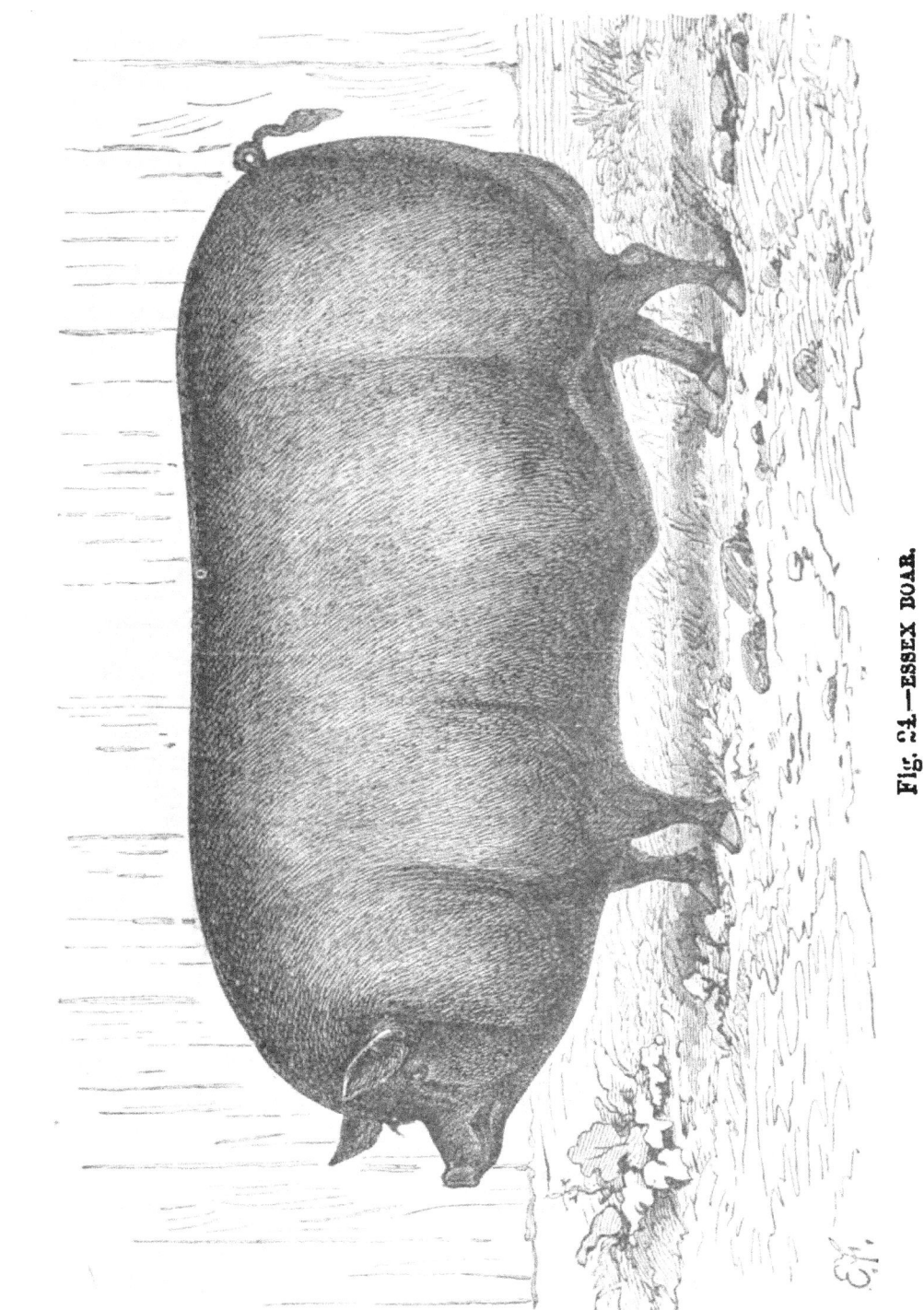

Fig. 24.—ESSEX BOAR.

Western's death, Mr. Hobbs purchased his best breeding sows. The difference between Lord Western's Essex and Mr. Fisher Hobbs' improved Essex, is shown very plainly by the two portraits which illustrate this section, the one drawn by Mr. Youatt, in 1845 (fig. 23), and the other from 'Emperor,' an eight-year-old working boar drawn for me in April, 1860 (fig. 22).

"The improved Essex, with symmetry, have more size and constitution than the original Essex-Neapolitans, and

Fig. 25.—ESSEX SOW.

this has been maintained without any crosses for more than twenty years, by judicious selection from the 'three distinct families.'"

Very excellent specimens of the Essex pigs are owned by various breeders in this country. We give engravings (figs. 24, 25, —) from photographs of animals owned by L. A. Chase, Esq., Northampton, Mass., descended from animals imported by Samuel Thorne, Esq., from Fisher Hobbs' stock. They are in only working condition.

IMPROVED OXFORDSHIRE.

"These black pigs," says Mr. Sidney, "although they are scarcely numerous enough to enable them to claim the title of a breed, are interesting, because representing a successful attempt to unite the best qualities of the Berkshire and improved Essex. The old Oxfordshire breed were very like the old Berkshire. The first great improvement is traced to two Neapolitan boars imported by the late Duke of Marlborough when Marquis of Blandford, and presented by him to Mr. Druce, senior, of Eynsham, and the late Mr. Smallbones, in 1837. These Neapolitans were used with Berkshire sows, some of which were the result of Chinese crosses. Two families of jet-black pigs were formed by Mr. Smallbones and Mr. Druce. On the death of Mr. Smallbones, Mr. Samuel Druce, jun., purchased the best of his stock, and had from his father, and also from Mr. Fisher Hobbs, improved Essex boars. The produce were a decided 'hit,' and very successful at local, Royal, and Smithfield Club shows. The improved Oxfords are of fair size, and all black, with a fair quantity of hair, very prolific, and good mothers and sucklers.

"Mr. Samuel Druce writes me: 'I have recently used one of Mr. Crisp's black Suffolk boars. In fact, wherever opportunity offers, I obtain good fresh blood of a suitable black breed, with the view of obtaining more lean meat than the Essex, better feeding qualities than the pure Berkshires, and plenty of constitution. I have never been troubled with any diseases among my pigs. Without change of boars of a different tribe, if of the same breed, constitution cannot be preserved. Where breeding in-and-in from a limited stock is persisted in, constitution is lost, the produce of each sow becomes small in size and few in number.' The Oxford dairy farms have a first-rate market for pork in the University. Porkers at thirteen to sixteen weeks are wanted to weigh 60 lbs. to 90 lbs.;

bacon pigs at nine to ten months, 220 lbs. to 280 lbs., but at that age the improved Oxfords are easily brought to 400 lbs."

BLACK AND RED PIGS.

"Birmingham has long been one of the greatest pig markets in the kingdom, and the pig-breeding of the district has been not a little affected and improved by the winter fat-stock show, which has for some years past been held there, at Bingley Hall, with great success. The town of Birmingham unites Staffordshire and Warwickshire. The old Warwickshire breed was a white or party-colored animal of the old-fashioned farm-yard type, and has never been improved into a special breed. The Staffordshire breed was the 'Tamworth.' At present the Tamworth are rapidly going out of favor with farmers, from the want of aptitude to fatten, and are being replaced by useful pigs, the result of miscellaneous crosses of no special character. The best are the middle-sized white pigs, a cross of the Cumberland-York with local white breeds, often called the Cheshire. The northern cross improves the constitution, and gives hair of the right quality, 'hard, but not too much or too coarse.'

"At Bingley Hall the class of Berkshire breeding-pigs under six months old generally brings from twenty to twenty-five pens. At present, however, the Berkshires in the Birmingham district are chiefly in the hands of amateur farmers, tenant farmers not having taken very kindly to them.

"But the breed must be spreading rapidly if the ready sale of the young pigs at the Birmingham show be taken as evidence.

"Mr. Joseph Smith, of Henley-in-Arden, one of the most successful exhibitors of Berkshires, keeps three or four sows, and sells all their young; and others find the demand for young pigs constant throughout the year.

"Mr. Thomas Wright, of Quarry House, Great Barr, (who did so much toward founding the Bingley Hall show,) considers the cross of the Berkshire with the Tamworth 'produces the most profitable bacon pigs in the kingdom, the Berkshire blood giving an extraordinary tendency to feed, and securing the early maturity in which alone the Tamworth breed is deficient. The cross of the Berkshire boar with large white sows has been found to produce most satisfactory results to plain farmers. My own notion with regard to all agricultural stock is, that we should abandon crosses and stick to our pure breeds, adapting them to our particular wants by careful selection.'

"The TAMWORTH BREED is a red, or red-and-black pig,—hardy, prolific, and the best specimens well shaped, but slow in maturing. It seems a near relation to the old Berkshire; but modern Berks breeders carefully exclude all red-marked pigs from their breeding-sheds. Reddish hairs at the tips of the ears of Essex would be permitted and admired. Mr. Alderman Baldwin, of Birmingham, is a noted breeder of this hardy, useful pig, which, however, does not seem to have any success as a prize winner. At the Royal Agricultural Show at Warwick, 1859, the Yorkshire and Berkshire breeds divided all the honors."

DEVONS.

"Devonshire," says Mr. Sidney, "has an excellent breed of black pigs, which partake, for the most part, of the character of the improved Essex and Berkshire. The climate seems to require less hair than the northern and midland counties. Mr. George Turner, the great cattle-breeder of Devon, has done a good deal in the last forty years towards improving the west country black pigs by his 'stud' and importations.

"The 'original Devon pigs were valued according to

the length of their bodies, ears, noses, tail, and hair; the longer the better, without reference to quality or substance,' just like some Devonshire squires of 500 ragged acres, who value themselves on the length of a pedigree unilluminated by a single illustrious name or action. 'They were of no particular color or character; but within the last forty years they have been improved perhaps more than any other stock, by judicious crosses and importations.' Within the last twenty years a good deal of Mr. Fisher Hobbs' stock (Essex) has been introduced, and seem well adapted to the climate. The Berkshires are also much approved. Mr. George Turner's stock 'are black, with short faces, thick bodies, small bone, and but little hair, and exhibit as much good breed, shape, and constitution, as any tribe of pigs in the kingdom, and have won as many prizes at the breeding-stock shows of the Royal Agricultural Society.'

"' At eighteen months old they generally make from 18 to 20 score—360 lbs to 400 lbs., sinking the offal.

"Some of the original breed of the county may still be seen in parts of North Devon; they will jump a fence that would puzzle many horses and some hunters. But, taken as a whole, the pig stock of Devonshire is far above the average of other counties; the black pig being, perhaps, the only foreigner who has ever been cordially welcomed as a settler in that very exclusive county."

DORSETS.

"Dorset," says Mr. Sidney, "has no reputation as a pig-breeding county; but one breeder, Mr. John Coate, of Hamoor, has achieved a reputation for his Improved Dorsets, by winning, amongst other prizes, the gold medal for the best pen of pigs in the Smithfield Club Show not less than five times, viz., 1850, 1851, 1852, 1855, and 1856.

"Mr. Coate writes me that he purchased, about 'twenty years ago, a boar and sow in Somersetshire, of a breed said to have been sent from Turkey. They resembled, in some measure, the wild boar,* being short on the leg, with very long, wiry hair, black in color, and very inclined to fatten. I was led to believe it was a mixture between the wild boar and Neapolitan breeds. I crossed them with some Chinese I had, and by so doing, *both ways*, produced the animals I named, when first exhibited, the 'Dorset breed,' although not properly; but they had, from their beauty, previously found their way into many farm-yards in the county. I had two distinct breeds to begin with (Mr. Coate means, I presume, the Chinese-Turks and the Turk-Chinese,) which I kept pure a long time for crossing; but as both wore away, have used my own stock as far akin as possible, and have once or twice introduced fresh blood by getting a boar as much like my own as I could. I have tried crosses with other breeds, but not liking the offspring, got rid of them again. Crosses answer well for profit to the dairyman, as you get more constitution and quicker growth; but for me, who sell a great number of pigs for breeding purposes, I find it will not do, as it requires many years to get anything like purity of blood again. With all animals, the first or second cross is good; but if you ever get away from the pure breed, it requires years and great attention to regain it, as the cross often shows itself in color or shape years after it has taken place, when you fancy you are quite safe.'

"There is no manner of doubt that Mr. Coate's Dorsets have been improved by a strong cross of Mr. Hobbs' improved Essex. Experienced pig judges tell me that they carry the relationship plainly in their faces; and this

---

* According to this description, they did not in the least resemble any wild boar I have ever seen.—S. S.

would be a safe cross, both being derived from Neapolitans.

"But Dorset, as a county, is so far from being celebrated for pigs, that one of the greatest dairy farmers, who feeds whole herds, writes me—'All I know is, that our breed of pigs is very bad.'

"They are, for the most part, black and white, of a Berkshire character. The ancient Dorset pig is said to have been blue, perhaps the original of the blue boar. One well-known parish in Dorset is called 'Toller Porcorum.'"

Mr. Sidney certainly deserves credit for the boldness with which he endeavors to classify the different breeds of English pigs. It is not an easy or an agreeable task.

It would seem from the facts given above that the White Breeds are decidedly of a mixed origin. The Yorkshire breeders furnish pedigrees, but if we may judge from the specimen given on page 67, these pedigrees, when analyzed, show conclusively that the breeders who have been most celebrated as prize-winners, have found it desirable to resort to an occasional cross. They have aimed to produce a pig that will grow rapidly, and fat at an early age. In other words, they have aimed, as breeders, to produce what we want as feeders. This is, we think, a mistake. The object of the breeder should be to produce a pig which, *when crossed with common sows*, will produce the best pigs for fattening.

Agricultural Societies will not allow a grade Shorthorn, or a grade Hereford, or a grade Devon, or a grade Ayrshire to compete with a thorough-bred. But both in England and America, pigs are shown without reference to pedigree; and as long as this is the case, the breeders of thorough-bred pigs receive injury rather than benefit from these exhibitions. None but thorough-breds should be allowed to compete with thorough-breds. The importance of "pedigree" is admitted, but the societies do not insist upon it, and the consequence is that nearly all the prizes go to grade pigs, or to some recently made-up breed.

If one of these successful exhibitors of a made-up breed is a conscientious man, he endeavors to keep his pigs pure, and every year they become more valuable for the purpose of improving common stock, but less likely to take a prize. Mr. Mangles' York-Cumberlands, of which we give a beautiful portrait on page 66, are as handsome pigs as can be desired; but if kept pure for a dozen generations, they will be no better than they are now for "show" purposes; in fact, they will probably not be as good. Some newly made-up breed, with equal refinement, but with stronger digestive organs, will take on fat more rapidly and will win the gold medal—as they themselves did when not half as valuable for the purpose of improving ordinary stock as they are now.

We cannot better conclude this account of the English breeds than by copying the following remarks from Mr. Sidney's book:

"It will be right to say a few words about two or three county pigs of no particular merit, but which, nevertheless, are 'familiar in our mouths as household words.' For instance, there is the HAMPSHIRE HOG—a name used, very unjustly, no doubt, to designate a county man as well as a county pig. There are some very pretty things to be said about the herds of swine in the New Forest, but they have been said so often, that they are scarcely worth repeating. The county animal is black or spotted with red, and about the size of a Berkshire, but coarser, and has had less attention paid to its improvement. There are also a considerable number of white pigs in Hampshire. Like every other breed within reach of a good market, they have been much improved within the last twenty years; but no Hampshire man has made himself celebrated as a pig-breeder, and I cannot find any instance of Hampshire pigs taking prizes at the Smithfield Show;

therefore, it may be concluded that, although the county abounds in useful animals, it is not worth while to resort to it either for establishing a new or improving an old breed. Of his class, the Berkshire is a better animal than the dark Hampshire hog, both having, when unimproved, a want of thickness through the shoulder, which has been corrected by a cross of Neapolitan or Essex, and both are slow feeders.

"The LINCOLNSHIRE PIG cannot now be distinguished from Yorkshire. At the Lincoln Royal Agricultural Society's Show, the prizes were easily carried away by Berkshires; but that proves nothing, as some judges never give a prize to a white pig, and others never to a black one.

"The SUFFOLK, a white pig, once appeared frequently in the catalogues, and in the prize-lists of the Smithfield Club Show, but of late years it seems to have given way to more popular names. Suffolk has a leading breeder of pigs in Mr. Crisp, of Butley Abbey; but he breeds both black pigs and white pigs, and calls his black pigs Suffolks, being a sort of cosmopolitan breeder, a purchaser of the best pigs he can find of any color. His most celebrated pigs are quite black. Mr. Barthropp, of Cretingham Rookery, celebrated for his Suffolk horses, but not a pig-breeder, writes of the swine of his native county in terms which might be applied to almost every district not distinguished by a thorough-bred sort. 'The old Suffolks were white, with rather long legs, long heads, flat sides, and a great deal of coarse hair; they made good bacon hogs, but were not so well adapted for porkers as the present improved Suffolks are. These are the white, with short heads, and long cylindrical bodies upon short legs, and fine hair, which breeders try to get long, fine, and thin. These are the best Suffolks; but there are a great many about the county, the result of crosses with the black Essex, which have 'no character,' although they are useful animals.' The best Suffolks, as before mentioned, are Yorkshire-Cumberlands, that have emigrated and settled in Suffolk, and thence been transported to Windsor.

"The NORFOLK PIG, also described by Youatt, is, according to the report of one of the best farmers in the county, 'an indescribable animal, the result of the mixture of many breeds in a *hocus pocus* or *porcus* style; and although they have improved of late years, the county stands very low in that division of live-stock.' 'They really are (writes another Norfolk farmer) a disgrace to our county. The only thing to recommend them is, that they are great breeders. If they would have three or four less, and better quality, it would pay better.' In the days of the first Earl of Leicester, he had, of course, some good pigs for the time, and they then found their way into book, and have remained there ever since. The only noted pig-breeder in Norfolk cultivates the improved Berkshire.

"BEDFORDSHIRE cannot boast of a county pig, but a pig was bred at Woburn, white, with occasional brown spots, and depicted in Youatt's original edition of this book, which I have the very best Bedfordshire authority for saying, was 'a good sort of pig, without any particular character, good feeders, but bad swillers, and they were therefore allowed to die out, and replaced by Berkshire sows, crossed with Suffolk boars. Indeed, the Bedfordshire breed were so little known, that a tenant of one of the first-class farms of that county told me that 'he did not know that they had a breed until he saw it marked over one of Prince Albert's pens, about ten years ago, at the Smithfield Club.'

"At present a white breed is the most fashionable, which means salable, in Bedfordshire.

"Another very eminent Bedfordshire farmer says: 'The

breed of pigs in this county is wretchedly bad, and has been ever since I have known it.'

"A third writes me: 'The Woburn breed, described by Youatt, was a good sort of pig, of no particular character, except great aptitude to fatten. They were discontinued, in consequence of the sows being very bad sucklers, in favor of a cross-bred animal, the produce of Berkshire sows and white Suffolk boars, the best that could be got. These are prolific, of good quality, can be fed at any age, and to a fair medium weight. A cross like this pays the farmer best.'

"Herefordshire has a useful white pig, but no attention has been paid to it.

"The dairymen in Cheshire breed and buy a great many dark pigs, black, black-spotted, and red-and-black, of the Shropshire and Welsh breeds, using Berkshire boars, and also Manchester or 'Yorkshire' boars.

"A tenant of R. Egerton Warburton, Esq., of Arley Hall, writes in answer to a set of questions which that gentleman was kind enough to circulate among his tenants:

"'There is no distinct Cheshire breed. The pigs are mostly cross-bred, short-eared, and long-sided. The favorite breed is a cross between Berkshire and Chinese.'

"The Shropshire, of which great numbers are introduced into Cheshire by traveling pig-jobbers, are of a dark red-and-black color, long-snouted, and lengthy; not very fine in the coat.

"The Welsh pigs are generally a yellow-white, but some are spotted black-and-white.

"The (Cheshire) dairymen depend more on these Welshmen and proud Salopians than on breeding. The cross of the Manchester boar with the Shropshire and Welsh produces a larger and coarser breed than the small Yorkshire.

"The Cheshire farmers buy in their stores at about sixteen weeks, feed them from eight to twelve months, and sell them weighing from 240 lbs. to 300 lbs. These are considered, in Cheshire, the best selling weights for bacon. I observe that the farmer who uses most Welsh pigs keeps them twelve months, and sells them at 300 lbs., which will scarcely pay for four months more keep than the Yorkshire, Manchester, and Shropshire sold after eight months.

"An immense improvement has taken place in Cheshire pigs within the last thirty years, in quality and weight. They are made fat at least six months sooner than thirty years ago.

"One farmer says few or no Irish pigs are brought into Cheshire; another, a good many, but not so many as formerly. The great importation is of Shropshire and Welsh. Yet a county member, who ought to be an authority, writes me that 'Shropshire cannot boast of a county pig.'

"As a general rule, dark pigs would seem to be in favor on English dairy farms.

"The MIDDLESEX is a name which has become known from winning prizes at the Smithfield Club, in 1841, 1848, 1850, 1851, 1854, 1856. It is not a county pig, but of the same class as the Windsor. Mr. Barber, of Slough, Buckinghamshire, is the principal breeder and exhibitor of Middlesex. Captain Gunter used to show it before he settled permanently in Yorkshire.

"The NOTTINGHAMSHIRE BREED, whatever that may be, has won one prize in Baker-street, and the Warwickshire crossed with Neapolitan two, many years ago.

FANCY BREEDS.

"By fancy breeds, I mean pigs named after a person or a place. The prizes awarded to pigs at the Smithfield Club Shows are a very good evidence that the breed, if a breed, had good feeding qualities, although it may not have been suited for the ordinary work and treatment of a farm. Cross-bred animals have had the greatest success. Pure Essex and Berkshire, and large Yorkshires, have not met as much success as at breeding stock shows. The most successful animals at Smithfield have been cross-bred. The prize-winning white pigs, under whatever name, have all had a large dash of Cumberland-York-Leicester; the black pigs, of Neapolitan-Essex.

"Among the most successful exhibitors at the Smithfield Club Shows, has been H.R.H., the Prince Consort, with what has lately been called *the Windsor breed.*

"This is a white pig, the result, apparently, of many crosses, the prevailing blood being small York-Cumberland. Thus, H.R.H. won, according to printed prize-list, in

    1846, with Bedfordshires.
    1847, " Bedfordshire and Yorks.
    1848, " Suffolks.
    1849, " Suffolks.
    1850, " Yorkshires.
    1851, " Bedfordshire and Suffolks.
    1852, " Suffolks.
  (These were, all but one, second prizes.)
    1853, " Suffolks.
 (First prize and gold medal for best pen of pigs in
      any class.)
    1854, " Windsors.

"And since that time only the breed has been called Windsors. His Royal Highness took a first prize in small boars at Warwick with his Windsor breed, and a commendation for a Berkshire sow.

"It is a tribe greatly in demand among gentlemen pig-breeders, and crosses admirably with strong county sows.

"The COLESHILL is a white pig, closely connected with

the York-Cumberlands bred at Coleshill, by the Earl of Radnor, who had stock from Earl Ducie, who had stock from Mr. Wyley, of Bransby, Yorkshire, and Mr. Brown, of Cumberland, for more than twenty years. The Coleshills, between 1847 and 1850, had great success at the Smithfield Club Shows; since that time, they seem to have somewhat lost their reputation, and two of my Yorkshire correspondents describe them as 'toys.' 'At one time they were of a good size, but they have by no means maintained the even character that would entitle them to the name of a breed." When any of Lord Radnor's stock pass into other hands in England, the produce generally ceases to be called Coleshills. They become Suffolks, Yorkshires, Middlesex, according to the fancy of the breeder. They are esteemed, and much better known among the fashionable pig-breeders in France than in England, and there their opponents term them 'drawing-room pigs'—(*cochons de salon*). The Coleshills carried off first prizes and gold medals at the Smithfield Shows in 1846 and 1847, and second prizes in 1844, 1845, 1847, and 1850.

"The BUSHEY BREED are white, bred by the wealthy banker, Mr. Majoribanks, and were long called Yorkshires, and have recently been named after their place of birth. They have no distinctive character to distinguish them from their competitors.

"The BUCKINGHAMSHIRE took the first Smithfield prize in 1840, but in these and many other names it is difficult to find any distinctive character."

This is additional evidence, if any were needed, that the most successful prize-winners resort to crossing. The whole system of awarding prizes to pigs needs a thorough revision. As it now stands, it is simply a means of enabling breeders to sell highly fed, cross-bred "toys" at high prices. The "Prince Albert Suffolks," which we now learn are nothing but high-bred grades, have been introduced into the United States. Perhaps the writer has less cause than he supposed, to regret that one which he kept until four years old, finally found her way to the pork barrel without ever breeding a single pig.

CHAPTER XI.

BREEDS OF PIGS IN THE UNITED STATES.

We have no "native" American pig. Our stock originally came from Europe, and principally from Great Britain. And it is highly probable that the largest and best specimens of the period were brought over by the colonists; and as improvements were afterwards effected in England, good animals of the improved breeds were imported.

Attempts have also been made to improve our pigs by using Chinese boars and their crosses; and there can be no doubt that individual breeders in this way succeeded in effecting a great improvement in the early maturity and fattening qualities of their stock. But although these attempts attracted considerable attention at the time, the pigs so obtained were never generally popular. They were too small and delicate for the prevailing taste of the period.

In 1832, the Improved Berkshires were introduced into the United States, and soon attracted the attention they so well deserved. In the course of half a dozen years, they were introduced into nearly every State in the Union. Breeders became excited. The agricultural papers were filled with communications extolling the merits of the Berkshires—and after a careful perusal of these articles at this time, we find that the statements were not as highly colored as might have been expected. As a rule, the pigs were quite as good as they were represented to be. It was hardly to be expected that breeders should say to intending purchasers, "It is of no use for you to get a well-bred pig unless you are prepared to give it better treatment than you do the common sort." The trouble was not in the pigs, but in the farmers. Berkshires were fully as valuable as the breeders claimed, and yet a great and wide-spread disappointment soon manifested itself. For a time the supply was not equal to the demand, and doubtless hundreds of pigs were sold as "pure Berkshires" that were nothing but grades. But the general complaint was that the Berkshires *were not large enough.* The advocates of the breed met this complaint by statements of weights, giving many instances where the Berkshires and their grades dressed 400 lbs. at a year old, and that at 18 or 20 months old, they could be made to weigh 500 or 550 lbs., dressed. One of the prominent breeders stated that he had a thorough-bred Berkshire that *gained* 496 lbs. in 166 days, and when killed, dressed 626 lbs.

To meet the demand for large pigs, fresh importations were made of the largest Berkshires that could be found in England. One boar, "Windsor Castle," imported in 1841, by Mr. A. B. Allen, it was claimed would weigh, at two years old, when in good flesh, 800 lbs. At the same time, Mr. Allen deprecated the prevailing taste for such large hogs, and very justly argued that smaller pigs, with less offal, would mature earlier and fatten more rapidly on a given amount of food. But then, as now, the demand was for the largest pigs that could be found, and it is said that this very boar was afterwards sold to a gentleman in Ohio for one thousand dollars.

But the excitement soon began to abate. Farmers who had paid $50, $100, and, in one instance we have met with, $250 for a single pair of Berkshires, found that their neighbors did not like the looks of the new comers. Ordinary pigs were selling at from $1 to $3 per cwt., and few could be persuaded to pay even $10 for a pair of thorough-breds. Thus ended the Berkshire excitement. The reaction was so great, that for years afterwards there were farmers who would not have received as a gift the best Berkshire in the world. And to this day, thousands who do not know a Berkshire pig when they see it, have a very decided prejudice against the breed.

A few years later, the Suffolks were introduced by the Messrs. Isaac & Josiah Stickney, of Boston. These gentlemen unquestionably procured the best specimens of the breed that could be purchased in England, and they bred them with great care and skill. Other importations were made, and the Suffolks have probably been more extensively diffused throughout the New England, Middle, and Western States than any other improved English breed.

About the same time, the Improved Essex were introduced, but, being entirely black, they never became popular in the Northern States. They are principally in the hands of our large stock breeders, and other gentlemen of wealth, but are rarely found on ordinary farms. Being in the hands of men knowing the value of pedigree, they are probably, to-day, as "*pure-bred*" pigs as can be found in the United States or in England.

The large Yorkshires were introduced soon after the breed became noted in England, and importations have been made from time to time. But no special efforts have been made to create an excitement in regard to this breed, and it has not been extensively diffused. The small Yorkshires, or Prince Albert Suffolks, were introduced about ten years ago, and, for a time, attracted considerable attention. But they are not favorites with the majority of farmers.

The above comprise the principal English breeds that have attracted any special attention in this country, and before alluding to breeds originating in the United States, it may be well to inquire *why* these valuable English breeds have never been favorites with the generality of our farmers?

That these breeds are not now, and never have been popular, is unquestionably a fact. Except some kept by the writer, we do not know of a single thorough-bred Berkshire, Essex, Suffolk, or Yorkshire pig within ten miles, and it is doubtful whether there are any in the county, although they have been repeatedly introduced. As a general rule, these thorough-bred pigs are kept only by persons who raise them to sell for breeding purposes. They are not kept for the sole object of making pork. For the latter purpose they are seldom as profitable as the offspring of a good common sow and a thorough-bred boar.

The handsomest pigs we have ever seen were so obtained; and one would think that farmers, seeing such a result, would continue to use thorough-bred boars. But such is seldom the case. They prefer to use one of these large handsome grades, rather than the smaller and more refined thorough-breds, and in this way the beneficial influence of the improved blood is soon lost.

We think this is the principal reason why these highly-refined English breeds are not favorites with ordinary farmers. Their real value consists in their perfection of form, smallness of bone and offal, and the great development of the ham, shoulder, cheeks, and other valuable parts; and added to this is their ability to transmit these qualities to their offspring. This ability is in proportion to their purity, and hence the value of pedigree. When one of these pure-bred boars is put to a good grade or common sow, we get precisely what we want—pigs having the form, the refinement, the early maturity, smallness of offal, and tendency to fatten of the thorough-bred, combined with the vigor, constitution, appetite, and great digestive powers of the larger and coarser sow. In other words, as far as the production of pork is concerned, we get a perfect pig—and there the improvement ends. We have attained our object, and all that we have to do is to repeat the process. To select boars from these grade pigs, and to use them in hopes of getting something better, is mere folly. It can lead to nothing but disappointment. And yet this is the common practice of those who are, once in a while, induced to try the thorough-breds. They soon find themselves possessed of a stock of non-descript pigs, none of them equal to the first cross, and some of them inferior to the sow first put to the thorough-bred boar. Then we hear complaints of the "degeneracy" of the improved breeds, when, in point of fact, no sensible man could expect any other result. Another cause of the unpopularity of the thorough-bred English pigs is, the wretched treatment to which they are often subjected. When we first commenced keeping thorough-bred pigs, a farmer of the neighborhood who, some years before, had paid a high price for a pair of Suffolk pigs, and who failed to raise a single thorough-bred pig from them, remarked, "You will soon get tired of this business. I have tried it. They won't breed. You are keeping them too fat. The only way to treat them is to turn them to a straw stack, and let them live on that." The fact that *he* never raised a pig from his sow did not commend his treatment, and we continued feeding our pigs sufficient food to keep them growing rapidly, and have had no cause to regret it. The only sow that has ever failed to breed with us was a Prince Albert Suffolk, purchased in the neighborhood from a farmer who had probably tried the "straw-stack" mode of feeding.

The aim of a good breeder of pigs is to get a breed that will grow rapidly and mature early. And the better the breed, the more rapidly will they grow. But the best stove in the world cannot give out heat without a supply of fuel; neither can the best-bred pig in the world grow rapidly without food; and the more thoroughly the power to grow rapidly has become established by long and careful breeding, the less capable does the pig become to stand starvation. It may sometimes be necessary to starve a pig for a short time when it has become too fat. In this case the pig gets food from its own fat and flesh, and sustains no permanent injury. But to starve a young, growing pig, is always injurious—and the more rapidly the pig is designed to grow, the more detrimental and permanent will be the effects of such treatment. The handsomest lot of white pigs we have ever raised, were from a sow got by a thorough-bred Earl of Sefton (Yorkshire) boar. She was a very large sow, and not coarse for her size. This sow we put to a thorough-bred highly refined Prince Albert Suffolk, and had a litter of " beauties." There was not a poor pig among them, and they were so uniform that it was difficult to tell one from another. The sow had been liberally fed, and at the time of pigging was very fat, and we continued to feed her and the little ones all they would eat. The result was a lot of pigs that we have never seen excelled. Encouraged by this result, we purchased from a neighbor, at an extra price, a litter of pigs got by the same thorough-bred boar, and at the same time another litter of common pigs from another neighbor. Both litters ran together, and had the same food and treatment, and the common pigs *did better than the grade Suffolks.*

The grade Suffolks were, in fact, decidedly poor pigs—a very different lot from the pigs from our own sow, got by the same boar. One cause of the difference must probably be assigned to the fact that the sow was not as large or as good as ours, and was not as well fed. And another reason for the difference was, *the pigs, for the first two months, had not had all the food they were capable of eating.* They never recovered from this neglect, and the common pigs were a stronger, more vigorous and healthier lot, and ultimately made much the heaviest pork. If we had had no other experience than this, we should certainly condemn thorough-bred pigs. But we *know* that the fault was not in the breed, but in the treatment which the sow and her young litter had received. Common pigs are better than improved pigs that have been injured, while young, by neglect and starvation, but the improved pigs, if the mother has been liberally fed, and they themselves are allowed as much food as they require to grow rapidly, will be found altogether superior to the common pigs, and vastly more profitable.

To say that, up to the time they shut them up to fatten, the majority of farmers half starve their pigs, will not be considered too strong an assertion by any one who has turned his attention to the subject. And this being the case, it is very evident that the improved English breeds cannot be popular—and the same is true of all other improved breeds of animals. We must adopt a better system of farming before we can hope to see the improved breeds of cattle, sheep, and pigs generally introduced and fully appreciated. Improved breeds necessitate improved farming, and improved farming cannot be very profitable without improved breeds, improved seeds, and improved implements. To tell a poor farmer that "it is just as easy to raise a good animal as a poor one," is telling him what, *in his case*, is not true. If he thinks he can do so merely by buying one or two improved animals to start with, he will soon find out his mistake. He should first improve his farm, and adopt a better system of feeding and management, and *then* he will find it nearly as easy to raise good animals as poor ones, and vastly more profitable.

We are now prepared to consider the breeds of pigs which are most popular in the United States, and may be able to discover the cause of their popularity.

CHESTER COUNTY WHITE PIGS. (Figure 27.)

The most popular and extensively known breed of pigs in the United States at this time is, unquestionably, the Chester County breed, or, as they are generally called, the "Chester Whites." The rearing and shipping of these pigs has become a very large and profitable business. One firm alone in Chester Co., Penn., informs us that, for the last three or four years, they have shipped from 2,500 to 2,900 of these pigs each year, and many other breeders have also distributed large numbers of them.

There are several reasons why the Chester Whites are

Fig. 27.—CHESTER COUNTY WHITE.

more popular than the English breeds. In the first place, they are a large, rather coarse, hardy breed, of good constitution, and well adapted to the system of management ordinarily adopted by the majority of our farmers. They are a capital sort of common *swine*, and it is certainly fortunate that they have been so extensively introduced into nearly all sections of the country. Wherever Chester Whites have been introduced, there will be found sows admirably suited to cross with the refined English breeds. No cross could be better than a Chester White sow and an Essex, Berkshire, or Small Yorkshire thorough-bred boar. We get the form, refinement, early maturity, and fattening qualities of the latter, combined with the strong digestive powers, hardiness, and vigorous growth of the Chester Whites. If the first cross does not give pigs possessing sufficient refinement and early maturity, a good, thrifty, well-formed sow should be selected from the litter and put to a thorough-bred boar, and this second cross will, so far as our experience goes, be as refined as is desirable for ordinary farm pigs. When the pigs are to be killed at four or five months old for fresh pork, a sow may be selected from this second cross, and again put to a thorough-bred boar. This is probably as far as it is desirable to carry the refining process. The pigs from this third cross would have $87\frac{1}{2}$ per cent of thorough-bred blood in them, and so far as the production of pork is concerned, would be more profitable than the thorough-breds.

We think this is the proper use to make of the Chester White pigs. They have many excellent qualities. They are large, hardy, strong, vigorous, have good constitutions, breed well, and are good mothers. Whether, as a breed, they are *thoroughly established*, is rather doubtful. There are probably families among them that have been bred long enough to permanently establish their good qualities. But it is certain that many Chester Whites have been sent out that produce litters, the pigs of which differ from each other as widely as the litters of common sows—and far more widely than the litter of a common sow put to a thorough-bred boar.

Paschall Morris, of Philadelphia, who has bred Chester Whites for many years, and who is thoroughly acquainted with the breed, describes them as follows: "They are generally recognized now as the best breed in this country, coming fully up to the requirements of a farmer's hog, and are rapidly superseding Suffolks, Berkshires, and other smaller breeds.

"The best specimens may be described as long and deep in the carcass, broad and straight on the back, short in the leg, full in the ham, full shoulder, well packed forward, admitting of no neck, very small proportionate head, short nose, dish face, broad between the eyes, moderate ear, thin skin, straight hair, a capacity for great size and to gain a pound per day until they are two years old. Add to these, quiet habits, and an easy taking on of fat, so as to admit of being slaughtered at almost any age, and we have, what is considered in Chester County, a carefully bred animal, and what is known elsewhere as a fine specimen of a breed called 'Chester County White.' They have reached weights of from 600 to 900 lbs.

"We have recently heard of a case where a farmer in the West had purchased some pigs from Chester County, and wrote back that part of them were full-blood, part half-blood, and part no Chesters at all. We know of another case where a purchaser insisted that a pig from Chester County was half Suffolk.

"There is considerable misapprehension about the Chester County breed, so-called. It is constantly forgotten that it is not an original, but a *made up* breed. They differ from each other quite as much as any one known breed differs from another. We have often seen them,— and the offspring, too, of good animals,—with long noses, which would root up an acre of ground in a very short time, slab-sided, long-legged, uneasy, restless feeders, resembling somewhat the so-called race-horse breed at the South, that will keep up with a horse all day on ordinary travel, and that will go *over* a fence instead of taking much trouble to go *through* it. They show more development of head than ham, and as many bristles as hair, and are as undesirable a hog as can well be picked up. Any traveler through Chester County can see such specimens continually. The standard of excellence in all animals, no matter how high or how pure may be the breed, so-called, is only to be *kept* up by judicious care in *feeding, breeding and management*. If either is neglected, they are sure to run out, and go down hill. With swine most especially, 'the breed is said to be in the trough.'

"When persons speak, therefore, of a pure Chester hog, or a half-blood, or a quarter-blood, we consider it only absurd. There is no such thing. By an original breed is meant, one that has been *long established*, and of which there are peculiar marks and qualities by which it has been long known, and which can be carried down by propagation. Such is the Devon cow and the Southdown sheep. The difference in results between an *original* and a recently made up breed may be compared to that between a seedling and grafted variety of fruit. If the seed of a very fine pear or apple is planted, there is no certainty, perhaps no probability, that the fruit will be the same as the parent. A graft of the parent tree, however, always produces the same. The results of the other are accidental. The law of breeding domestic animals, that 'like produces like,' applies more certainly to distinct and original breeds, like Devons or Southdowns, than to a made up breed of recent origin, like the Chester County hog. The owner of a very fine animal, who, for several years, has been selecting his stock carefully, and feeding them liberally, has the *chances* greatly in his favor that 'like will produce like,' but there are very often to be seen very poor specimens from good parentage, and also very good individual animals from very inferior parents. We have one now which, at a year old, will not weigh over 250 lbs.; she is the offspring of large and well-shaped parents. In adjoining pens are others which, at the same age, will weigh about 400 lbs. The hair, sometimes, is straight, at others, waved or curly. The ear is often small and erect, then again *large*, thick, and lopped, like that of an elephant. Blue spots often appear on the skin, and sometimes black spots on the hair. These and other great variations, in external form and other qualities show that the Chester County pig represents his *individual self*, and is not a *type* of a well established *breed*.

"In the best specimens there are, however, a contribution of more valuable points than belong to any other. As Ellman and Webb and Bakewell did with sheep, and with a far less favorable starting point, it is hoped some one may be found to take up the Chester County hog, and, by a *persevering course* of careful selections, breed him up to a still higher standard, and give him a more definite type and character.

"Any one can do this for himself, but the constant variations in their appearance would seem to show that it has not *yet* been done by any one. An impure Southdown lamb cannot be produced from a full-bred dam and sire; and yet a misshapen and ill-shaped pig is sometimes produced from what are called 'pure Chesters.'"

Coming from a distinguished advocate and breeder of Chester County pigs, this statement is as candid as it is explicit. We may take it for granted that the Chester Whites are *not* an established breed, like the Berkshires or Essex. They will not breed true. This would not be so very objectionable in itself, but it follows that, when we wish to improve our common stock, we should not resort to a Chester County boar. It is an axiom in breeding that we should use nothing but thorough-bred males. Chester County *sows*, when judiciously selected, are far superior to our ordinary run of pigs, and this breed will long continue valuable for the purpose of furnishing good breeding sows to cross with some good thorough-bred boar of the English breeds.

And it may be, as Mr. Morris suggests, that we shall

be able to so improve the Chester County pigs by such "a persevering course of careful selections," as to give the breed a better and "more definite type and character," and to so thoroughly establish these characters, that we may use the boars with a reasonable prospect of improving any common breed with which they are crossed. Until this is done, however, it will be a mistake to use Chester County boars, *except for the purpose of obtaining large, vigorous sows, to be crossed with some thoroughly established breed.*

The "Hog Breeders' Manual," a little work published in the interest of Chester County pigs, says: "The Chester and Suffolk make a very fine cross. If a new breed could be made by crossing these two breeds, and continuing, and the offspring were a uniform mixture of the two, I should consider it the maximum of perfection."

In other words, the Chester Whites are too coarse, and need to be refined by crossing with some of the thorough-bred English breeds. This is undoubtedly true; and coming from a prominent breeder of Chester Whites, may be regarded as decisive on this point. But why should a farmer wish for a "new breed" when, by using a thorough-bred Suffolk boar on a Chester White sow, he can attain at one step the "maximum of perfection?" True, he cannot breed from these perfect pigs. He cannot hope to make them "more perfect;" but, by continuing to use thorough-bred boars, he is always sure of obtaining good pigs. What more is needed? We think it would be a mistake if the Chester White breeders should refine their pigs too much. The chief value of the breed consists in its size and vigor, and in furnishing strong, healthy sows, to be crossed with thorough-bred boars of a refined breed. There is no object to be gained by refining, or, in other words, reducing the size and vigor, of the Chester Whites.

THE "CHESHIRE," OR JEFFERSON COUNTY PIGS. (Fig. 28.)

This is a breed of pigs originating in Jefferson County, N. Y. For a dozen years or more they have been exhibited at the Fairs of the N. Y. State Agricultural Society, and for the last six or seven years have carried off nearly all the prizes offered for pigs of the large breed. They were first exhibited, to the best of our recollection, under the names of "Cheshire and Yorkshire," and afterwards as "Improved Cheshires," and in 1868, one of the largest breeders exhibited them as "Improved Yorkshires."

Fig. 28.—JEFFERSON COUNTY PIG.

These different names, in different years, indicate the nature of the breed. They have been very extensively distributed throughout the country, and especially in the West, under the name of "Cheshires." It would be better, we think, to call them the "Jefferson County" pigs, as indicating the place rather than the nature of their origin. The latter is uncertain, while there can be no doubt that Jefferson County is entitled to the credit of establishing a very popular and valuable breed of pigs.

The old Cheshire pig was one of the largest and coarsest breeds in England, but Sidney says "these unprofitable giants are now almost extinct." A Cheshire (England) correspondent of this author writes, under date March 17, 1860, as follows: "The old gigantic, long-legged, long-eared pig, of a large patched black and white color, is all but extinct. My son met with a fine specimen last year in a sow which he brought to breed with our boar of the Berkshire small breed, but changed his mind and fed her. She showed no propensity for fattening at two years old. She weighed, when killed, 42 score, 12 lbs.—852 lbs; but as 3¹|₄ d. per pound was the best we could get for her, we took her for the family, and the meat was surprisingly good. She was lean fleshed. The hams weighed 77 lbs each."

It is said that a large sow of the old Cheshire breed was taken from Albany to Jefferson County, and about the same time some thorough-bred Yorkshires were introduced into the same neighborhood from England. We have not been able to definitely establish the fact, but it is highly probable that the pigs which were first exhibited at the N. Y. State Fair as "Cheshire and Yorkshire" were from Yorkshire boars, crossed with the descendants of this sow. The pigs, as we recollect them when first exhibited, were very large, rather coarse, but well shaped. Since then, they have, from year to year, approximated more closely to the Yorkshires. They are still large, but have finer bones and ears. The best specimens, as shown by the leading breeders, are as handsome pigs as can be desired. Color, white; small, fine ears, short snout, with a well-developed cheek; long and square bodied; good shoulders and hams, and very small bones for such large hogs.

As compared with the Chester County breed, they are nearly or quite as large, have finer bones, ears, and snout, and are altogether superior in form, beauty, and refinement to any Chester Whites we have ever happened to see. They have doubtless obtained this refinement from the Yorkshires. The leading breeders in Jefferson County admit very freely that the breed is of mixed origin, but it is claimed that they have been kept pure sufficiently long to thoroughly establish the breed. We believe that this, at any rate, has been the aim of some of the breeders. When thoroughly established, the breed will occupy a similar position to pure-bred large Yorkshires. The boars will be useful to cross with coarse Chester White sows, where larger hogs are desired than can be obtained by using Berkshire, Essex, or Suffolk boars.

THE MAGIE (OHIO) PIGS.

The Hon. John M. Millikin, in his Prize Essay on the Agriculture of Butler County, Ohio, gives an account of a large breed of pigs which have obtained considerable celebrity in some parts of the West. He says:

"No county in the United States, of equal area, has produced so many hogs of a superior quality as the county of Butler. The breed which is so highly esteemed by our farmers is the result of careful and judicious breeding, conducted by our best breeders in this county, and the adjoining county of Warren, for the last forty years.

"The precise history of the method adopted to produce this popular breed of hogs cannot be given as fully and as reliably as its present value and importance demand. The best information, of a reliable character, which can be obtained, gives us to understand that as early as about 1820, some hogs of an improved breed were obtained and crossed upon the then prevailing stock of the county. Among the supposed improved breeds of hogs, there were the Poland and Byefield. They are represented as being exceedingly large hogs, of great length, coarse bone, and deficient in fattening qualities. Subsequently more desirable qualities were sought for, and the stock produced by the crosses with Poland, Byefield, and other breeds, underwent very valuable modifications by being bred with an esteemed breed of hogs then becoming known, and which were called the Big China. They possessed important qualities in which the other breeds were sadly deficient. At a later period, Mr. Wm. Neff, of Cincinnati, an extensive pork packer, and fond of fine cattle and hogs, made some importations of fine stock from England. Among them were some Irish Graziers. They were white in color, of fair size, fine in the bone, and possessing admirable fattening qualities. Berkshires, about the same time, were attracting much attention, and both breeds were freely crossed with the then existing stock of the county. The result of these crosses was highly advantageous in the formation of a hog of the most desirable qualities. The Berkshires had obtained, with many breeders, great favor, while others objected to them, because they thought them too short, and too thick in the shoulder. Nevertheless the Berkshire blood was liberally infused into our stock of hogs, but in such a judicious manner, as to obviate the objections urged against them, and to secure their conceded good qualities.

"Since the formation period of our breed of hogs, as above stated, there have been no material or decided innovation upon the breed thus obtained. Our breeders have carefully selected and judiciously bred from the best animals thus produced among us. Where defective points have been apparent, they have been changed by careful breeding. There has been, for many years, no admixture of any other breed of hogs. Our own breed is now, and has been for nearly thirty years, the stock predominant in this county. Our breeders believe that they have a *well established breed* of hogs, which is unsurpassed in the most desirable qualities of a good hog. This breed of hogs, although of recent origin, may be regarded as thoroughly and permanently established. They have been bred so long, and with such judgment and uniform success, that they may be confidently relied upon as possessing such an identity and fixity of character as a distinct breed, as to give assurance that they will certainly and unmistakably propagate and extend their good qualities.

"They can scarcely be said to have a well-established, distinctive name. They are extensively known as the 'Magie stock.' They are sometimes called the 'Gregory Creek hogs,' but more generally they are known as the 'Butler County stock.' It will be doing no one injustice to say that D. M. Magie has bred these hogs as extensively and judiciously as any other man in the county. He has not only bred them for his own use, but also to supply the extensive demand that has been made upon him from all parts of the West and North-west.

"While we claim that Butler County has more good hogs than any other county in the State, we do not desire to do our neighbors any injustice by appropriating all the credit for this breed of hogs to ourselves. Warren County assisted in the formation and establishment of this breed of hogs. They continue to raise them in their purity and full perfection, and take into the market as fine lots of hogs as have ever been raised and sold.

"In verification of what we claim, we propose to show the averages of hogs sold and delivered to packers—not isolated cases, nor single specimen hogs, but the lots of hogs raised by our farmers, and sold in the market. These hogs are usually wintered over one winter, and are sold at ages ranging from eighteen to twenty-one months. Mr. David M. Magie has made the following sales:

One lot of 63 Hogs..........Average weight ....444 lbs.
"   "   40 "   ....   "   "   ....417 "
"   "   80 "   ....   "   "   ....433 "
"   "   60 "   ....   "   "   ....400 "
"   "   72 "   ....   "   "   ....413 "
"   "   100 "   ....   "   "   ....408 "
"   "   43 "   ....   "   "   ....467 "
"   "   35 "   ....   "   "   ....451 "
"   "   120 "   ....   "   "   ....458 "

Thomas L. Reeves sold 39 head, 17¼ months old, averaging 459 lbs.
Jeremiah Beaty " 35 " " 438⅜ "
L. Miltenberger " 35 " " 449 "
Abraham Moore " 40 " " 466 "
William Gallager " 71 " " 473 "
" " the first 22 of same, " 528 "

"These are individual lots, among many which have been noticed as remarkable for their high average. Although they have never been equaled, so far as the public know, yet some may regard another kind of evidence as more conclusive. To such we submit the following facts, kindly furnished by Mr. Chenoweth, who, for many years, has weighed the hogs packed by Jones & Co., at Middletown, in this county. The hogs there packed are mainly furnished by citizens of this county, and Warren County.

In the season of 1862–3, there were packed 4,956 hogs, averaging 305 lbs.
" " 1863–4, " " 5,538 " " 276 "
" " 1864–5, " " 5,370 " " 282 "
" " 1865–6, " " 6,003 " " 345 "
" " 1866–7, " " 5,013 " " 335 "

In 1867–8, a dozen of the best lots averaged 459 lbs.

"These figures," says Mr. Millikin, "must decide the superiority of our breed of hogs over all others. To produce such averages, the stock must be of the best quality, and then care and judgment in breeding must be practiced, and good attention given in raising and fattening."

It is evident that the Butler County farmers know how to raise and fatten hogs. But it does not follow, from the figures given above, that there is necessarily any special merit in the Magie breed. We know farmers who take great pride in having heavy hogs, who make them weigh from 450 to 500 lbs. at 18 or 20 months old. And yet these very hogs are of such a kind, that no intelligent man, who is acquainted with the merits of the improved breeds and their grades, would tolerate on his farm for any other purpose except to cross with some highly refined thorough-bred boar. We are not acquainted with the Magie hogs, and would not be understood as saying that they are of this kind. They may be the best breed in the world, but the fact that the credit of the breed is awarded to the county, and not to individuals, does not indicate any special and decided characteristics. Breeds do not originate in this way. It is not to the farmers of Leicestershire that we owe the Leicester sheep, but to Robert Bakewell; it is not to the farmers of Durham, but to the Messrs. Collins, that we owe the Durham or Shorthorn cattle. The farmers of Sussex are entitled to no credit for the Sussex or Southdown sheep. Ellman did more to improve these sheep than all the other Sussex farmers had accomplished in a thousand years. We owe the Essex hogs to Lord Western and Fisher Hobbs, and not to the farmers of the county—and so it always is. The old Essex pig was one of the worst in England; Fisher Hobbs made it one of the very best in the world.

CHAPTER XII.

EXPERIMENTS IN PIG FEEDING.

Boussingault weighed a litter of five pigs at the moment of their birth. The smallest weighed $2^1/_2$ lbs., and the largest $3^1/_2$ lbs., the average of the whole litter being $2^3/_4$ lbs. each. At the end of 36 days he weighed them again, and they then averaged 17.3 lbs., showing a gain of nearly 3 lbs. each per week. During the next five weeks they gained $3^1/_2$ lbs. each per week.

The quantity of food consumed was not ascertained.

Dr. M. Miles, Professor of Agriculture in the Michigan Agricultural College, has made some valuable experiments in feeding young pigs, in which the amount of food consumed and the gain each week were accurately ascertained.

Six grade Essex pigs, two weeks old, were selected for the experiment. They weighed 25 lbs., or a little over 4 lbs. each. At the end of the first week they weighed $46^1/_2$ lbs., showing a gain of a little over $3^1/_2$ lbs. each,—a gain of about 90 per cent in one week. At the end of the second week they weighed 84 lbs. They were then divided into two separate pens, three in a pen. The pigs in pen A weighed $43^1/_2$ lbs., and those in pen B $40^1/_2$ lbs. At the end of the third week the three pigs in pen A weighed $52^1/_2$ lbs.; those in pen B, 54 lbs. At the end of the fourth week pen A weighed $66^1/_2$ lbs.; pen B, $69^1/_2$ lbs. At the end of the fifth week pen A weighed 79 lbs.; pen B, $85^1/_2$ lbs. Sixth week, pen A, $89^1/_4$ lbs; pen B, $93^1/_4$ lbs.

At this time one of the pigs in pen B met with an accident and was killed. It weighed, alive, 30 lbs., and dressed, 23 lbs.

To the end of the eighth week the pigs were allowed all the new milk they would drink, and what corn they would eat in addition. After the eighth week the milk was discontinued, and they were allowed all the corn-meal they would eat, mixed fresh with a little water.

During the first week the pigs consumed about $23\frac{1}{2}$ lbs. each of milk, and gained $3\frac{1}{2}$ lbs. each.

Second week, they consumed 48 lbs. each of milk, and gained a little over 6 lbs. each.

Third week, consumed 47 lbs. milk, and gained $3\frac{3}{4}$ lbs. each.

Fourth week, consumed 52 lbs. milk, and gained 5 lbs. each.

The amount of food consumed for each pound of live weight of the pigs was—

1st week.	2d week.	3d week.	4th week.
3.93 lbs.	4.42 lbs.	2.95 lbs.	2.57 lbs.

The gain for each hundred pounds of live weight was—

1st week.	2d week.	3d week.	4th week.
86.00 lbs.	80.64 lbs.	26.78 lbs.	27.69 lbs.

The amount of food consumed to produce one pound of increase was—

1st week.	2d week.	3d week.	4th week.
6.53 lbs.	7.70 lbs.	12.52 lbs.	10.56 lbs.

These experiments, confirmed as they are by others giving similar results, show conclusively that a young animal eats much more, in proportion to live weight, than an older one. Thus, for each pound of live weight, the pigs ate nearly 4 lbs. of milk the first week, and only $2\frac{1}{2}$ lbs. the fourth week. It would also seem that the younger the animal, the more rapidly it gains in proportion to the food consumed. Thus, it required about 7 lbs. of milk the first fortnight to produce a pound of increase, and $11\frac{1}{2}$ lbs. the second fortnight, or about 65 per cent more.

So far, therefore, these results strikingly confirm the conclusion we should arrive at from theoretical considerations, that the more food an animal can eat, digest, and assimilate in proportion to its size, the more it will gain in proportion to the food consumed.

During the second month, each pig ate, in pen A, 37 lbs. of milk per week, and 1 lb. each of oats and corn, and gained 2.83 lbs. each per week. This also shows a great falling off in the consumption of food in proportion to live weight, and a still greater falling off in the rapidity of increase in proportion to the food consumed.

During the eighth week it required nearly double the amount of food to produce a pound of increase as during the fourth week.

After the eighth week, as we have said, the pigs were fed exclusively on corn-meal. The following table shows the amount of food consumed by each pig per week, and the increased growth obtained from it.

	Food consumed by each pig per week.		Increase in live weight of each pig per week.		Food required to produce 1 lb. of increase.	
	Pen A.	Pen B.	Pen A.	Pen B.	Pen A.	Pen B.
3d month	8.00 lbs.	24.56 lbs.	1.70 lbs.	6.56 lbs.	4.68 lbs.	3.81 lbs.
4th "	13.75 "	18.25 "	3.50 "	4.50 "	3.92 "	4.06 "
5th "	16.00 "	25.00 "	4.25 "	5.93 "	3.82 "	4.22 "
6th "	16.66 "	25.87 "	0.75 "	4.62 "	8.83 "	5.24 "
7th "	Meal. 14.16 " Roots. 5.00 "	23.18 "	2.04 "	3.75 "	{ 7.00 " 2.42 "	5.98 "
Average of 5 months.	Meal....14.23 " Roots... 1.00 "	23.39 "	2.67 "	5.14 lbs.	Meal..5.32 " Roots.0.37 "	4.55 "

It should be remembered that these pigs were all of one litter, and that in both pens they had the same food, (except that during the seventh month of the experiment the pigs in Pen A were allowed roots in addition to the corn meal) were fed at the same time, and in the same conditions, and both were allowed all they would eat, and yet the pigs in pen B ate 61 per cent more food than those in pen A, and gained over 92 per cent more.

We cannot tell why one pig differs from another pig of the same litter. But, aside from this, it is not difficult to understand why pigs, that eat more food, should gain more in proportion to the food consumed. It is owing to the fact that, in the case of the small eaters, nearly all the food is used merely to support the vital functions. In a previous chapter (page 21) we have endeavored to explain this matter in detail.

One of the pigs in pen A gained nearly as much as those in pen B; and had the pigs been fed separately, the result would doubtless have been even more strikingly in favor of the large eaters.

The following table shows the weight of the pigs when six weeks old (the fourth week of the experiment), and for each month afterwards.

		Weight of each pig at 6 weeks old. lbs.	Weight at 10 weeks old. lbs.	Weight at 14 weeks old. lbs.	Weight at 18 weeks old. lbs.	Weight at 22 weeks old. lbs.	Weight at 26 weeks old. lbs.	Weight at 30 weeks old. lbs.	Gain in 24 weeks. lbs.	Gain in 20 weeks. lbs.
Pen A.	Pig 1	21	31½	36	42	53	52	59	38	27½
	" 2	23	33¼	38	49½	60	64	69	46	25¾
	" 3	22½	35¾	47	71½	101	120½	133	110½	97¼
Pen B.	" 4	23½	48½	80	100½	131	148½	156	132½	108½
	" 5	23	44	64	79½	96½	118½	142	129	98

At ten weeks old, the pigs were not allowed any more milk, but were allowed all the corn-meal they would eat. From this time, until they were 30 weeks old, a period of five months, pig 1 gained 27½ lbs., pig 2, 35¾ lbs., and pig 3, 97¼ lbs., all in the same pen. Taking the pens together, we have shown that the pigs in pen A ate about 5½ lbs. of food to produce 1 lb. of increase, while the pigs in pen B required only 4½ lbs. to produce the same result. But it is undoubtedly true that these figures do not show the whole advantage to be gained by having pigs that can eat and assimilate a large amount of food. Pig 3 probably ate much more than his proportion of the food, and gained even still more in proportion to the food consumed. Thanks to Professor Miles, we are not left wholly to conjecture on this important point. Finding that pigs No. 1, and No. 2 had no tendency to lay on fat, and that they were increasing only in bone and muscle, he thought it desirable to ascertain the amount of food which each pig consumed; so, at the beginning of the 21st week of the experiment, the pigs were put in separate pens, and allowed, as before, all the food they would eat. During the first week afterwards,

>Pig 1 ate 11 lbs. meal.
>" 2 " 12½ " "
>" 3 " 25½ " "

During the month the pigs ate and gained as follows:

>Pig 1 ate 48½ lbs. meal, and lost 1 lb.
>" 2 " 51½ " " " gained 4 lbs.
>" 3 " 100 " " " gained 19½ lbs.

Pigs 1 and 2, together, ate precisely the same amount of food as pig 3 alone. But in the one case, the 100 lbs. of corn gave $19^1/_2$ lbs. of increase, and in the other only 3 lbs. So much for a good appetite.

CHAPTER XIII.

LAWES AND GILBERT'S EXPERIMENTS IN PIG FEEDING.

The most extensive experiments on fattening pigs are those made by J. B. Lawes, Esq., and Dr. J. H. Gilbert, at Rothamstead, near St. Albans, in England. These gentlemen have, for many years, devoted themselves to such investigations; their experiments were conducted with the greatest care, and in the most thorough manner, and the results are worthy of entire confidence. Unfortunately, as it seems to us, the experiments were confined exclusively to pigs shut up to fatten; and no particular attention was given to the breed, or the previous history of the pigs. The principal object of the experiments was to ascertain the best kinds of food for fattening pigs, and the best proportion of nitrogenous to non-nitrogenous food.

"In the selection of the animals," say Messrs. Lawes & Gilbert, "it was only sought to get such as resembled each other in character, age, and weight, in the several pens; and, with this view, a competent person was employed to go to the various sties and markets in the neighborhood to purchase animals suited to our object.

"Forty pigs were purchased, as nearly as possible of the same character, and all supposed to be about nine or ten months old. The pigs were weighed and marked, and thirty-six of them selected out, and divided into twelve lots, of three each, in such a manner as to give *equal weights* in each lot, but it was found that, in selecting them by weight alone, 'they did not secure animals of sufficiently equal feeding quality in the several pens.' On the following day, therefore, they were changed from pen to pen, so as to provide, as much as possible, a similarity in this respect between pen and pen, and, at the same time, to retain a near equality in weight also. This being done, the weights stood as follows:

TABLE I.—SHOWING THE WEIGHTS OF THE PIGS WHEN ALLOTTED TO THE PENS, FEB. 3, 1856.

Nos. of the Pigs.	Pen 1—lbs.	Pen 2—lbs.	Pen 3—lbs.	Pen 4—lbs.	Pen 5—lbs.	Pen 6—lbs.	Pen 7—lbs.	Pen 8—lbs.	Pen 9—lbs.	Pen 10—lbs.	Pen 11—lbs.	Pen 12—lbs.
1..................	146	146	142	142	140	133	133	132	130	129	131	130
2..................	121	122	115	123	123	123	124	133	124	128	128	115
3..................	112	112	113	113	115	122	121	117	119	120	120	129
Total weight of 3 Pigs.	379	380	370	378	378	378	378	382	373	377	379	374

"The allotment thus completed, all the pigs were fed on a mixture of one part bean meal, one part lentil meal, two parts Indian corn-meal, and four parts bran—these being the foods fixed upon for the subsequent experiment. The pigs were allowed as much of this food as they would eat." "Upon this mixture," say the experimenters, "all were kept for twelve days, prior to commencing the exact experiment, in order that they might become accustomed to their new situation, and reconciled to their new companions, for, in the allotment, the various purchases had necessarily been intermixed—in some cases, greatly to the disapprobation and discomfort of the individuals of those purchases. For a time, constant quarrels ensued, and the molested animals frequently jumped from pen to pen, until they fell in with former associates. Indeed, at first, it was no uncommon occurrence, after they had been left for some time, to find some pens almost deserted, and others crowded. The use of the whip was found to be very efficacious in settling these disputes, and at length, all seeming to live amicably together, the exact experiment was commenced on Feb. 14, twelve days after the first allotment."

This account will prove interesting, and furnish valuable hints to such of our agricultural colleges as may contemplate making experiments on animals. It shows, furthermore, taken in connection with the weight of the pigs, that little attention had been paid to their breeding or management. They were, evidently, common store hogs, active, and quarrelsome, and the fact that, at from nine to ten months old, they only weighed 112 to 146 lbs., indicates that the farmers of England do not treat their pigs much better than the farmers of America.

During this preliminary period of twelve days, the pigs gained as follows:

TABLE II.—SHOWING THE GAIN OF EACH PIG DURING THE TWELVE DAYS OF THE PRELIMINARY PERIOD.

Nos. of the Pigs.	Pen 1—lbs.	Pen 2—lbs.	Pen 3—lbs.	Pen 4—lbs.	Pen 5—lbs.	Pen 6—lbs.	Pen 7—lbs.	Pen 8—lbs.	Pen 9—lbs.	Pen 10—lbs.	Pen 11—lbs.	Pen 12—lbs.
1	30	11	21	31	28	24	15	13	26	20	6	19
2	14	20	16	8	5	21	2	11	18	10	22	15
3	17	11	15	10	20	22	20	26	11	9	16	21
Total gain in each Pen	61	42	52	49	53	67	37	50	55	39	44	55

During this period, of twelve days, the pigs were all fed on the same food, and were allowed all they chose to eat, and yet it will be seen that the gain is far from uniform. "Those pigs," say the experimenters, "having flourished most, which had fallen in for the lion's share, whilst the weaker ones, which had been obliged to sulk in the rear until their more powerful companions had indulged to the full, clearly indicated their misfortunes by their weights. After that time, however, very little irregularity occurred from this cause—vigilant care being taken that each animal should have his full share of food—and it soon happened that the mere approach of the whip, was sufficient to awe the pugnacious delinquent into humble retreat, while his weaker neighbor, in his turn, took precedence at the trough. These ill-tempers, though at first very troublesome, gave way surprisingly by a little perseverance, and the evil of them, in the course of comparative experiments is, after all, much less than in submitting to a faulty allotment."

The experiment proper, commenced Feb. 14, and continued eight weeks. The following table shows the weight of each pig at the commencement of the experiment:

TABLE III.—SHOWING THE WEIGHT OF EACH PIG AT THE COMMENCEMENT OF THE EXPERIMENT, FEB. 14.

Nos. of the Pigs.	Pen 1—lbs.	Pen 2—lbs.	Pen 3—lbs.	Pen 4—lbs.	Pen 5—lbs.	Pen 6—lbs.	Pen 7—lbs.	Pen 8—lbs.	Pen 9—lbs.	Pen 10—lbs.	Pen 11—lbs.	Pen 12—lbs.
1	176	157	163	173	168	157	148	145	156	149	137	149
2	135	142	131	131	128	144	126	144	142	138	150	130
3	129	123	128	123	135	144	141	143	130	129	136	150
Total weight of 3 Pigs.	440	422	422	427	431	445	415	432	428	416	423	429

The following table shows the weight of the pigs at the end of the experiment, after being fed eight weeks:

TABLE IV.—SHOWING THE WEIGHT OF EACH PIG AT THE END OF THE EXPERIMENT.

Nos. of the Pigs.	Pen 1—lbs.	Pen 2—lbs.	Pen 3—lbs.	Pen 4—lbs.	Pen 5—lbs.	Pen 6—lbs.	Pen 7—lbs.	Pen 8—lbs.	Pen 9—lbs.	Pen 10—lbs.	Pen 11—lbs.	Pen 12—lbs.
1	279	293	239	298	264	263	249	287	205	176	197	263
2	229	237	183	198	182	230	191	243	148	182	198	175
3	235	228	200	183	206	250	284	249	175	172	206	245
Total weight of 3 Pigs.	743	758	622	679	652	743	724	779	528	530	601	683

The food selected for the experiment was a mixture—1st, bean and lentil meal; 2d, Indian corn-meal, and 3d, bran.

As beans and lentils are, at present, little used as food for pigs in the United States, we shall not be far wrong in considering them as equivalent to peas. The object of the experiment was not merely to ascertain which of these foods was most nutritious, but what is the best proportion of feeding them. Accordingly, each of the pens had an unlimited allowance of some one of these three classes of foods, some of them having no other food, except in the case of bran, while others were allowed a restricted quantity. Thus:

Pen 1—was allowed a mixture of equal parts of bean and lentil meal *ad libitum*.

Pen 2—2 lbs. per pig, per day, of Indian corn-meal, and bean and lentil mixture *ad libitum*.

Pen 3—2 lbs. of bran per pig, per day, and bean and lentil mixture *ad libitum*.

Pen 4—2 lbs. of Indian meal, 2 lbs. of bran per pig, per day, and bean and lentil mixture *ad libitum*.

Pen 5—Indian corn-meal *ad libitum*.

Pen 6—2 lbs. of bean and lentil mixture, and Indian meal *ad libitum*.

Pen 7—2 lbs. bran, and Indian meal *ad libitum*.

Pen 8—2 lbs. of bean and lentil mixture, 2 lbs. bran, and Indian meal *ad libitum*.

Pen 9—2 lbs. of bean and lentil mixture per pig, per day, and bran *ad libitum*.

Pen 10—2 lbs. of Indian meal per pig, per day, and bran *ad libitum*.

Pen 11—2 lbs. of bean and lentil, 2 lbs. of Indian meal per pig, per day, and bran *ad libitum*.

Pen 12—Bean and lentil mixture, Indian corn-meal, and bran, each separately, and *ad libitum*.

The results ought to afford answers to the following questions:

Are peas (bean and lentil) as good, or better, than Indian corn, for fattening pigs?

Is it better to feed them alone, or mixed together, and in what proportions?

What is the value of bran as a food for fattening pigs, in conjunction with peas or Indian corn, or both?

When pigs are allowed all they will eat of peas, Indian corn, and bran, how much of each will they eat, and in what proportions?

The following table shows the gain of each pig during the experimental period of eight weeks:

TABLE V.—SHOWING THE GAIN OF EACH PIG DURING THE EXPERIMENTAL PERIOD OF EIGHT WEEKS.

Nos. of the Pigs.	Pen 1—lbs.	Pen 2—lbs.	Pen 3—lbs.	Pen 4—lbs.	Pen 5—lbs.	Pen 6—lbs.	Pen 7—lbs.	Pen 8—lbs.	Pen 9—lbs.	Pen 10—lbs.	Pen 11—lbs.	Pen 12—lbs.
1	103	136	76	125	96	106	101	142	49	27	60	114
2	94	95	52	67	54	86	65	99	6	44	48	45
3	106	105	72	60	71	106	143	106	45	43	70	95
Total gain of 3 Pigs	303	336	200	252	221	298	309	347	100	114	178	254

The pigs making the greatest gain are those in pen 8, which had 2 lbs. of peas (beans and lentils), and 2 lbs. of bran each, per day, and all the Indian corn-meal they would eat in addition. These pigs gained $14^1/_2$ lbs. each, per week, or over 2 lbs. per day. The next best gain is in pen 2, with 2 lbs. of Indian meal each, per day, and all the pea meal they would eat. They gained exactly 2 lbs. per day.

With Indian meal alone, the pigs gained not quite 9¼ lbs. each, per week. With Indian meal, and a small allowance (2 lbs. each, per day,) of peas, the gain is not quite 12½ lbs. per week; while with Indian meal, and 2 lbs. each, per day, of bran, the gain is over 12¾ lbs. per week. The most curious result, however, is in pen 12, where the pigs had all they would eat of each of the three kinds of food. The gain is but a fraction over 10½ lbs. each, per week.

The following table shows the amount of food consumed by each pig, per week, and the average increase in live weight, per head, per week, during the experimental period of eight weeks:

TABLE VI.—SHOWING THE AVERAGE WEEKLY CONSUMPTION OF FOOD, AND THE INCREASE, PER HEAD, DURING THE TOTAL PERIOD OF THE EXPERIMENT.

Nos. of Pens.	DESCRIPTION AND AVERAGE QUANTITIES OF FOOD CONSUMED PER PIG, PER WEEK, IN LBS.			*Average increase per week, per pig, in lbs.*
	Limited food.	*Ad libitum food.*	*Total food consumed per week, per pig, lbs.*	
1....	None.	63 lbs. bean and lentil meal,	63	12.62
2....	14 lbs. Indian meal.	52 " " " " "	66	14.00
3....	14 lbs. bran.	40¼ " " " " "	54¼	8.33
4....	14 lbs. Indian meal, & 14 lbs. bran.	31½ " " " " "	59½	10.50
5....	None.	45¼ lbs. Indian meal.	45¼	9.21
6....	14 lbs. bean and lentil meal.	44¼ " " "	58¼	12.42
7...	14 lbs. bran.	44¼ " " "	58¼	12.87
8....	14 lbs. bean and lentil meal, and 14 lbs. bran.	36¾ " " "	64¾	14.46
9....	19¼ lbs. bean and lentil meal.	18 lbs. bran.	37¼	4.16
10....	19¼ lbs. Indian meal.	23½ " "	42¾	4.75
11....	14 lbs. bean and lentil meal, and 14 lbs. Indian meal.	18 " "	46	7.42
12....	None.	28½ lbs. bean&lentil meal. 25½ lbs. Indian meal. 3 lbs. bran.	57	10.58

It is very evident that bran, fed in a large quantity, or with a small proportion of other food, is a very indifferent food for pigs. It is too *bulky*, in proportion to the nutriment it contains. The pigs were weighed every two weeks, and it was so obvious after the first weighing, that the pigs in pens 9 and 10 were not getting food enough (though having all the bran they would eat), that the limited food was increased to 3 lbs. per pig, per day, instead of 2 lbs. But, even with this addition, it is clear that the pigs did not get sufficient nutriment. Their stomachs were not capable of holding enough of this bulky, and probably rather indigestible, food.

The pigs in pens 9 and 11, ate precisely the same amount of bran per week, but the pigs in pen 11 were allowed $8^3/_4$ lbs. of meal more than pen 9; and it will be seen that this $8^3/_4$ lbs. of extra meal produced over $3^1/_4$ lbs. of extra increase.

Comparing pen 1 with pen 5, it will be seen that the pigs having pea meal alone, gain over 3 lbs. a week more than those having Indian meal alone; but the pigs in pen 1 ate more pea meal than the pigs in pen 5 did of Indian meal, and the actual increase from the food consumed is, if anything, rather in favor of the Indian meal. It will be found that 100 lbs. of pea meal produce 20 lbs. of increase, while 100 lbs. of Indian meal produced 20.3 lbs. increase. It would seem from this that 100 lbs. of peas will not produce any more pork than 100 lbs of corn. At the same time, it would seem that pigs will grow or fatten faster on peas than on corn. They are capable of eating more peas than corn.

By comparing pens 2 and 6, we have the same general indications. In pen 2, the pigs had pea meal *ad libitum*, and 2 lbs. of corn meal each, per day; while in pen 6, they had Indian meal *ad libitum*, and 2 lbs. of pea meal each, per day. Pen 2 ate the most food, and gained the most rapidly. But still the amount of food required to produce a given increase is almost identical. In pen 2, 100 lbs. of meal produced 21.2 lbs. of increase; in pen 6, 21.3 lbs.

The more we study these results, the more are we impressed with the importance of the study of physiology and breeding, in connection with the chemistry of food. Thus, in the same pen, on the same food, one pig gains 45 lbs., and another 114. In another pen, one gains 65, and another, on the same food, 143 lbs. And so it is in all our experiments on animals. There is a cause for this, and we cannot but hope that the subject will receive more attention from scientific investigators than they have hitherto bestowed upon it.

We should remark that, in pen 5, with Indian meal alone, one of the Pigs, No. 1, during the first fortnight, gained over 2 lbs. per day, while the other two only gained about half as much. Before the end of the first fortnight, however, "it was observed that this fast gaining pig, and one of the others, namely, No. 3, had large swellings on the side of their necks, and that, at the same time, their breathing had become labored.

"It was obvious," say Messrs. Lawes and Gilbert, "that the Indian corn-meal alone was, in some way, a defective diet; and it occurred to us that it was comparatively poor, both in nitrogen, and in mineral matter, though we were inclined to suspect that it was a deficiency of the latter, rather than of the former, that was the cause of the ill effects produced. We were, at any rate, unwilling, so far to disturb the plan of the experiments, as to increase the supply of nitrogenous constituents in the food, and accordingly determined to continue the food as before, but at least to try the effect of putting within reach of the pigs a trough of some mineral substances, of which they could take if they were disposed. The mixture which was prepared was as follows:

20 lbs. finely sifted coal ashes,
4 lbs. common salt,
1 lb. superphosphate of lime.

"A trough containing this mineral mixture was put into the pen at the commencement of the second fortnight, and the pigs soon began to lick it with evident relish. From this time the swellings, or tumors, as well as the difficulty in breathing, which probably arose from pressure of the former, began to diminish rapidly. Indeed, at the end of this second fortnight, the swellings were very much reduced, and at the end of the third fortnight, they had disappeared entirely.

The three pigs consumed of the mineral mixture, described above, 9 lbs. during the first fortnight, 6 lbs. during the second, and 9 lbs. during the third.

It may be also well to state that 'a butcher, with a practised eye, selected and purchased the carcass of one of these [Indian corn fed] pigs, which had been diseased, from among the whole thirty-six, after they had been killed and hung up.'"

Messrs. Lawes and Gilbert also made a second series of experiments on 36 pigs, divided as before, into 12 pens. The foods used were the same as in the first series, except that barley-meal was substituted for Indian corn, and the pigs were allowed 3 lbs. each, per day, instead of 2 lbs.

The pigs were about nine months old, and ranged from 105 lbs. to 138 lbs. each. They were shut up in the pens April 26, and allowed all they would eat of a mixture of equal parts of bean and lentil meal, barley-meal, and bran. They were kept on this food until May 9, when they were again weighed, and the exact experiment commenced. All the pigs seem to have done remarkably well on this food, many of them gaining over 2 lbs. a day.

During the subsequent experimental period, however, no less than five of the pigs died, and for this reason we will not enter into a detailed account of the experiment.

The five pigs that died were in five different pens, feeding on different food. But it appears that they all belonged to one of the purchased lots of eight, and possibly to one litter, and, as Messrs. Lawes and Gilbert remark, "the loss was probably due to the bad constitution of the animals." The weather, however, was very hot, and unfavorable to the health of pigs kept closely confined and fed on rich food.

The gain of some of the pigs in this series was quite remarkable. Thus, in pen 2, which was allowed 3 lbs. of barley-meal per pig, per day, and bean and lentil meal *ad libitum*, one of the pigs gained 120 lbs. in eight weeks, or 15 lbs. a week. In the same pen, the other two pigs gained, one 65 lbs., and the other 99 lbs., during the same period, and on the same food. In pen 5, with barley-meal alone, *ad libitum*, one of the pigs gained 142 lbs. in the eight weeks, or $17^3/_4$ lbs. a week. One of the other pigs in this pen gained 87 lbs., and the other pig died.

It is very evident from these experiments that the success of a pig-feeder will depend much more on good judgment in selecting, or on care in breeding, the pigs he intends to fatten, than on the particular kind of grain given to them.

The best result of any pen in this series was where the pigs were allowed a mixture of 1 part bran, 2 parts bean and lentil meal (say pea-meal), and 3 parts barley-meal. The three pigs on this food gained 310 lbs. in eight weeks, or within two pounds of 13 lbs. each per week. Another pen, having precisely the same food, gave almost exactly the same gain, or 307 lbs. in eight weeks. An adjoining pen, having the same food, but a greater proportion of bean and lentil meal, and less barley-meal, gained 283 lbs. in the eight weeks, or about $11^3/_4$ lbs. each per week. One hundred pounds of the former mixture gave 20 lbs. of increase; of the latter, $18^1/_4$ lbs.

Messrs. Lawes and Gilbert also made a third series of experiments, the pigs being fed on dried codfish, in conjunction with bran, Indian meal, bean and lentil meal, and barley-meal, in different proportions. The codfish was boiled in water, and the meal mixed with it before being fed to the pigs.

The following table shows the composition of this dried codfish, together with the composition of the other foods used in this and the preceding experiments:

TABLE SHOWING THE COMPOSITION OF THE DIFFERENT KINDS OF FOOD USED IN MESSRS. LAWES' AND GILBERT'S EXPERIMENTS ON PIGS.

Description of Food.	Dry Matter.		Ash.		Nitrogen.		Fatty Matter.	
	Inclusive of Ash.	Organic Matter only.	In Fresh Substance.	In Dry Matter.	In Fresh Substance.	In Dry Matter.	In Fresh Substance.	In Dry Matter.
Egyptian Beans	88.30	83.57	4.73	5.35	4.24	4.80	2.29	2.60
Lentils—Lot 1	87.30	82.43	4.87	5.58	4.52	5.18	2.23	2.55
" Lot 2	86.62	81.64	4.98	5.75	4.56	5.26	2.21	2.55
Indian Meal—Lot 1	89.70	88.33	1.37	1.53	1.72	1.92	5.10	5.68
" " Lot 2	89.89	88.62	1.28	1.42	1.95	2.17	5.59	6.22
Bran	84.79	78.77	6.02	7.10	2.61	3.08	4.92	5.80
Barley—Lot 1	82.38	80.19	2.19	2.66	1.82	2.21	2.34	2.84
" Lot 2	80.95	78.77	2.18	2.69	1.83	2.26	2.33	2.88
" Lot 3	82.53	80.48	2.05	2.48	1.55	1.88	1.41	1.71
Dried New Foundland Codfish	59.26	40.60	18.66	31.49	6.60	11.13	0.90	1.52

In pen 1 the pigs were given, and compelled to eat, 14 lbs. each of codfish, per week, mixed with equal parts bran and Indian meal. Of this mixture they had all they could eat, and consumed 47 lbs. each, per week, and gained 10.09 lbs. each.

In pen 2, each pig had, as above, 14 lbs. codfish, and ate with it 45¼ lbs. Indian meal alone, per week, and gained 12.15 lbs.

In pen 3 the pigs had a mixture of equal parts Indian meal and bran, and as much codfish as they chose to eat. They ate 47 lbs. of the mixture of bran and meal, and only 7$^1/_2$ lbs. of codfish each, per week, and gained 8.94 lbs.

It will be seen that, when left to their own choice, the pigs in pen 3 ate only about half as much codfish as those in pens 1 and 2, where their other food was kept back until they had eaten their allowance of 2 lbs. of codfish per day.

The above pigs were about nine or ten months old, and were similar in character, weight, etc., to the pigs in the first two series of experiments.

In another series of experiments with eight pigs, seven months old, and "more finely framed" than the preceding pigs, 1 lb. of codfish was given to each pig, per day, with, in one pen, barley-meal alone, and in the other, with a mixture of 2 parts barley-meal, and 1 part bran.

In pen 4, the pigs ate 7 lbs. of codfish, and 49 lbs. of bran and barley meal each, per week, and gained 9.40 lbs.

In pen 5, the pigs ate 7 lbs. of codfish, and 57$^1/_2$ lbs. of barley-meal each, per week, and gained 11.75 lbs.

These facts will prove interesting and useful to farmers living near the sea-shore, in localities where fish are used for expressing oil, and where the refuse is sold for manure, or for food for pigs. An analysis of this refuse, taken in connection with the above experiments, should indicate its value as food for pigs, and it is an easy matter to calculate the value of the manure made by the pigs.

CHAPTER XIV.

SUGAR AS FOOD FOR PIGS.

Messrs. Lawes & Gilbert also made some experiments on pigs to ascertain the nutritive value of sugar as compared with starch.

Twelve pigs weighing from 72 lbs. to 98 lbs. each were placed in four pens, 3 pigs in a pen. Lentils and bran were selected as the nitrogenous food, and in pens 1, 2 and 3 the pigs were allowed 3 lbs. of lentil meal, and 1 lb. of bran each per day, and in addition, the pigs in pen one were allowed all the sugar they would eat, and those in pen 2, all the starch they would eat, and in pen 3 a mixture of equal parts starch and sugar. The pigs in pen 4 were furnished, in separate troughs, all the lentil meal, bran, starch and sugar they would eat. The experiment was continued 10 weeks. In pen 1, the pigs ate nearly 2 lbs. of sugar each per day, and in pen 2, a nearly identical quantity of starch; the other food being the same in kind and quality in both pens. The increase obtained from 100 lbs. of food was in pen 1, 20.8 lbs., and in pen 2, 19.9 lbs.

The pigs in pen 3, having a mixture of equal parts starch and sugar, and the same quantity of lentil meal and bran as in pens 1 and 2, ate 2¾ lbs. each per day of the starch and sugar. The increase from 100 lbs. of total food was 19.8 lbs.

In pen 4, where the pigs were allowed all they chose to eat of the different foods, each pig ate per day on the average, lentil meal 4 lbs. 6 oz., bran 3½ oz., starch 3¾ oz. and sugar 2 lbs. 2 oz. They ate more food and gained more rapidly than in any other pen. The increase from 100 lbs. of food was 21.3 lbs.

Without going into further details, it is evident that the pigs show a great preference for sugar as compared with starch, but it does not appear that sugar produces any materially greater increase than starch. Certainly there is no benefit approximating in the slightest degree to the increased cost of sugar; and it is very doubtful whether we should gain any marked advantage by converting our barley into malt or of growing sugar beets instead of ordinary beets or mangel wurzel.

We should add that a mixture of 20 lbs. of coal and wood-ashes, 2½ lbs. of superphosphate of lime and 2½ lbs. of common salt was placed in troughs in the pens. This quantity being distributed to the 12 pigs during each period of two weeks.

Messrs. Lawes & Gilbert say: This mineral mixture was always taken with the greatest avidity and relish; so much so, that the animals would leave their other troughs the moment the fresh supply of this was put within their reach. They were, moreover, upon the whole, very healthy throughout the experiment, and yielded good rates of increase.

In Messrs. Lawes' & Gilbert's account of these experiments, the actual gain of each pig is not given. But since writing the above, we have found the weights of the pigs at the commencement and at the end of the experiment, from which it appears that

the pigs in pen 1 gained 8.2 lbs. each per week.
" " " 2 " 8.2 lbs. " " "
" " " 3 " 9.1 lbs. " " "
" " " 4 " 10.4 lbs. " " "

To a practical farmer these actual figures are more interesting than mere percentage results. From this it would appear that, leaving the question of cost and profit out of the question, there may be cases where, with an unlimited supply of other food, a little sugar may be given to a pig with advantage. A pig with a delicate appetite might be given ordinary food, and then when he had eaten all he would of it, by mixing a little sugar with the food, he might be induced to eat more.

CHAPTER XV.

THE VALUE OF PIG MANURE.

There is much misconception in regard to the relative value of manure from different animals. It is often said that the manure of pigs is richer than that from cattle, horses, or sheep. This is sometimes the case, and sometimes not. It depends entirely on the food. An animal does not "make manure" any more than a stove makes ashes, or a thrashing machine makes grain, chaff, and straw. We feed a thrashing machine with a certain number of bundles of wheat, and get from it a certain amount of grain, straw, and chaff—but the machine does not make them. It was all in the bundles, and the machine merely separates them. And so it is in the case of an animal. A pig has no more to do in making rich or poor manure than a thrashing machine has in making white or red wheat. It depends entirely on the food.

There is little or no difference in the composition or value between the manure of a pig fed on clover, and that of a sheep, or a cow, or a horse, fed on clover. But if a pig is fed on clover, and the sheep is fed on straw, the manure of the pig will be by far the most valuable, simply because the clover contains a greater proportion of the more important elements of plant-food.

A ton of corn, fed to a pig, will not give manure worth as much as a ton of clover hay fed to a sheep, for the simple reason that a ton of clover hay contains more of the valuable constituents of plant-food than a ton of corn. But a ton of pig manure from a corn-fed pig may be, and often is, worth more than a ton of sheep manure from sheep fed on clover hay. The explanation of these apparently contradictory statements is this: A ton of corn contains more nutritious matter than a ton of clover. It contains more starch and oil, and these are digested and assimilated by the pig, and consequently there is a less quantity of matter to be voided as excrements. On the other hand, although a ton of clover contains a greater proportion of the more valuable elements of plant-food than a ton of corn, yet the clover does not contain nearly as much nutritious food as the corn. There is a large proportion of crude material that cannot be digested, and this is voided in the excrements; consequently, we get more manure from the ton of clover hay than from a ton of corn. It is not worth as much, weight for weight, but it is worth more as a whole, because there is more of it. In other words, a ton of pig manure from corn may be worth as much again, as a ton of sheep manure from clover hay; and, in point of fact, pig manure is ordinarily worth much more *per ton* than the manure from cows, horses, or sheep. But, at the same time, it is equally true that, if the same food was fed to a sheep that we feed to the pig, the manure of the sheep would be equally valuable. Pig manure is usually more valuable, in proportion to its weight or bulk, than ordinary farm-yard manure, because the pigs are fed on more nutritious food, or, in other words, on food containing a less proportion of crude, indigestible matter, and consequently we get less bulk of manure from the pig, but it is more valuable. But it is a grave error to suppose that a pig will make better manure than a sheep, a cow, or a horse.

The following table, prepared by Mr. Lawes, shows the average composition of different articles of food, together with the relative value of the manure made from the consumption of one ton of each food.

	Total dry matter.	PER CENT.			Value of manure in $ & cts. from 1 ton (2,000 lbs) of food.
		Total mineral matter (ash).	Phosphoric acid reckoned as phosphate of lime.	Potash.	Nitrogen.	
1. Linseed cake	88.0	7.00	4.92	1.65	4.75	19.72
2. Cotton-seed cake*	89.0	8.00	7.00	3.12	6.50	27.86
3. Rape cake	89.0	8.00	5.75	1.76	5.00	21.01
4. Linseed	90.0	4.00	3.38	1.37	3.80	15.65
5. Beans	84.0	3.00	2.20	1.27	4.00	15.75
6. Peas	84.5	2.40	1.84	0.96	3.40	13.38
7. Tares	84.0	2.00	1.63	0.66	4.20	16.75
8. Lentils	88.0	3.00	1.89	0.96	4.30	16.51
9. Malt dust	94.0	8.50	5.23	2.12	4.20	18.21
10. Locust beans	85.0	1.75			1.25	4.81
11. Indian meal	88.0	1.30	1.13	0.35	1.80	6.65
12. Wheat	85.0	1.70	1.87	0.50	1.80	7.08
13. Barley	84.0	2.20	1.35	0.55	1.65	6.32
14. Malt	95.0	2.60	1.60	0.65	1.70	6.65
15. Oats	86.0	2.85	1.17	0.50	2.00	7.70
16. Fine pollard†	86.0	5.60	6.44	1.46	2.60	13.53
17. Coarse pollard‡	86.0	6.20	7.52	1.49	2.58	14.36
18. Wheat bran	86.0	6.60	7.95	1.45	2.55	14.59
19. Clover hay	84.0	7.50	1.25	1.30	2.50	9.64
20. Meadow hay	84.0	6.00	0.88	1.50	1.50	6.43
21. Bean straw	82.5	5.55	0.90	1.11	0.90	3.87
22. Pea straw	82.0	5.95	0.85	0.89		3.74
23. Wheat straw	84.0	5.00	0.55	0.65	0.60	2.68
24. Barley straw	85.0	4.50	0.37	0.63	0.50	2.25
25. Oat straw	83.0	5.50	0.48	0.93	0.60	2.90
26. Mangel wurzel	12.5	1.00	0.09	0.25	0.25	1.07
27. Swedish turnips	11.0	.68	0.13	0.15	0.22	91
28. Common turnips	8.0	.68	0.11	0.29	0.18	86
29. Potatoes	24.0	1.00	0.32	0.43	0.35	1.50
30. Carrots	13.5	.70	0.13	0.23	0.20	80
31. Parsnips	15.0	1.00	0.42	0.36	0.22	1.14

* The manure from a ton of undecorticated cotton-seed cake is worth $15.74; that from a ton of cotton-seed, after being ground and sifted, is worth $13.25. The grinding and sifting, in Mr. Lawes experiments, removed about 8 per cent of husk and cotton. Cotton-seed, so treated, proved to be a very rich and economical food.

† Middlings, Canielle.   ‡ Shipstuff.

This table is of great value to the farmer. Hitherto, we have worked pretty much in the dark in regard to the profit or loss of fattening pigs. Many farmers contend that there is no profit in feeding hogs, while others claim that, when the manure is taken into consideration, there is no farm stock that pays so well. But it must be confessed that the wildest estimates are often made in regard to the value of the manure. By the aid of the above table it will not be difficult to form a pretty correct estimate of the value of the manure from any given lot of pigs, provided the kind and amount of food consumed is known.

Thus, if a pig was fed exclusively on corn from the time it was weaned until it had gained 350 lbs., it would eat about 1,500 lbs. of corn. Now, as the manure from a ton of corn is worth $6.65, the manure from 1,500 lbs. is worth $4.99. We may assume, therefore, that when pigs are fed on corn, in the production of every hundred pounds of pork, live weight, we get $1.42 worth of manure. Or, assuming that a fat pig will dress 80 per cent of its live weight, we may conclude that, in the production of every hundred pounds of pork, we get manure worth $1.78. In other words, in calculating the profit or loss of feeding pigs on corn, we may add $1^3/_4$ cents per pound (in gold), to the price of the pork for the value of the manure obtained.

On the other hand, if the pigs are fed on peas, we get manure worth more than twice as much, and may add $3^1/_2$ cents a pound to the price of the pork for the value of the manure made in its production. In this case, if pork sells for 7 cents per pound, we may calculate that for every dollar's worth of pork sold, we have 50 cents' worth of manure; or, if the pork sells for $10^1/_2$ cents per pound, for every dollar's worth of pork sold we have 33 cents' worth of manure in the pig pen.

Boussingault states that pigs from 5 to 6 months old will eat 19 lbs. of green clover per day—equal to about 5 lbs. of clover hay each. On such food we may safely calculate that a good pig will gain half a pound of *pork* a day; and if so, a pig that would dress 200 lbs. would have eaten green clover equal to one ton of clover hay;

and as the manure from a ton of clover hay is worth $9.64, we may calculate that every hundred pounds of pork so produced, leaves us $4.82 worth of manure.

When pigs are fed skimmed milk, we shall probably not be far wrong in estimating that the manure made in producing 100 lbs. of pork is worth $5.00.

Taking these four estimates together, and striking a mean, we have the following result:

Value of manure in producing 100 lbs. of pork from Indian corn............$1.78
" " " " " " Peas.................. 3.56
" " " " " " Clover... ...... :...... 4.82
" " " " " " Skimmed Milk........ 5.00

    Average of the whole.. $3.79

In other words, where pigs are fed on clover and skimmed milk during the summer, and are afterwards fattened on half peas and half corn, we may calculate that every pound of pork sold, leaves on the farm $3^1|_4$ cents' worth of manure.

It must be borne in mind that these are *gold* prices, and also that this is merely the value of the manure made by the pigs from the food consumed. The litter and other materials thrown into the pen have not been taken into the account. The pig cannot be credited with the manure so obtained. If we throw into the pen 100 lbs. of pea or bean straw, we add about 19 cents to the value of the manure heap; but this is not derived from the pig, but from the straw; and so it is with anything else thrown into the pen. The pig converts it into manure, but adds nothing to its value. The pig creates nothing. Whatever of value there is in the manure heap is derived from the food consumed, and from the materials used as litter. And yet it is nevertheless true, that we can obtain from the pig pen a large amount of valuable manure that otherwise would be wasted.

On farms, we have seldom time to attend to such matters, and there is not as great a necessity for it; but persons who have only a garden or small place, should have a pig pen, with a small yard attached, into which all the refuse material of the garden can be conveniently thrown—such as the clippings of the lawn, weeds, potato tops, pea and bean haulm, leaves, coal ashes, the loose dirt that is raked up in the garden beds, alleys, and walks, and the thousand and one things that we denominate rubbish. The whole of it should go into the yard attached to the pig pen. This is a much better way of disposing of it, than endeavoring to make a "compost heap." With such a yard, there never need be any trouble in determining where the materials in the wheel-barrow should be emptied. You have always a place for all rubbish that is lying around loose, and it will be an easy matter to keep the premises neat and clean.

"But oh, the smell!" exclaims a gentleman who let his Irish coachman keep a pig, "I cannot endure it." True; but this is the fault of the man, and not of the pig. A respectable, well brought up pig is the cleanest of all our domestic animals. Let him be washed once a week, and let plenty of dry earth, or soil of any kind, be scattered freely and frequently about the pen and yard, and all trouble from this source will cease, and the pig, if well bred, and well fed, will become one of the most popular features of the establishment, and he will be profitable also. He will pay in using up the refuse from the house and from the garden; pay in delicious hams, spare-ribs, and tenderloin; pay in firm, white, sweet lard; and, above all, he will pay in furnishing a large, rich compost for the garden, which, with the addition of a little superphosphate and guano, will pay double and treble in the abundance of crisp vegetables and well developed fruit.

The main point in managing a pig pen in such a case is, to furnish an abundance of earth to keep it clean. The pigs will root it over and mix it with the manure. The earth, especially if of a sandy nature, will at once favor decomposition and absorb the gases, and they in their turn will develop the plant-food in the soil, and we get a large quantity of manure that is free from smell, and not unpleasant to work over or use in the garden.

Where horses are kept, the refuse litter from the stables should be thrown into the pig pen. Horse manure is apt to ferment too rapidly, while pig manure is very sluggish. Mixing the two together improves both; and besides, the horse manure, when dry, makes a good bed for the pigs, and saves litter.

For garden vegetables, rich manure is especially valuable. It is desirable to concentrate the manure as much as possible. We do this by fermentation, which reduces the bulk, and at the same time renders the plant-food in the manure more immediately available. The plan here suggested, of throwing the dry manure from the horse stables into the pig pen, will tend still more to concentrate the manure. Pigs void large quantities of liquid, which contains nearly all the nitrogen of the food. The horse manure will absorb this, and, of course, we get a much more concentrated manure from the pig pen than when straw alone is used for bedding. We may not get *any more* plant-food from the two combined than we should if the droppings from the stable and from the pig pen were used separately, but we get it in a more concentrated form and in a more available condition; and this is a point of far greater importance than is usually supposed. We are inclined to believe that many of the diseases which affect vegetables in our old gardens are caused, or at least increased, by the excessive accumulation of carbonaceous matter in the soil, caused by the frequent use of manure deficient in phosphoric acid, potash and ammonia. The manure from a pig pen littered with horse droppings, thoroughly decomposed and mixed with earth, would furnish garden vegetables with all the plant-food they required in an available condition, and there would be less danger from fungi than where a large quantity of poor manure was frequently used.

On many farms half the value of the manure made by pigs is wasted. There is no part of the establishment so miserably managed as the pig pen. It is often nothing more than a pen of rails, with a little hovel in one corner, covered with corn-stalks, or straw, and the pigs are left to eat the corn on the ground, and wallow in mud and filth. If pork can be made at a profit in this way, it must be a good business when conducted properly.

CHAPTER XVI.

PIGGERIES AND PIG PENS.

In selecting a site for a pig pen, the first requisite is dryness. A side hill, sloping towards the barn-yard, is a desirable location; and if this cannot be found in a convenient place, it is not a difficult or expensive matter, with a dirt scraper and a span of horses, to form a basin in the barn-yard, using the dirt to make a high and dry foundation for the pig pens, and forming a slope towards the basin, so that the liquid from the pens will rapidly drain away to the manure heap. If the soil is not dry, it must be drained with tile, or stone underdrains, at least two feet deep; and if there is sufficient fall, four feet would be far better. These underdrains are not designed to carry off the water from the surface, but to make the soil underneath dry. Surface drainage must be attended to also; for, as the liquid from well-fed pigs is the most valuable portion of the manure, it is especially important that the whole of it should either be absorbed by the straw or other bedding in the pen, or drain away to the manure heap.

The next important consideration in locating the pig

pen is, convenience of feeding. Where there is much milk or whey, the pen should be located with reference to conveying it to pigs with the least labor. The only objection to having a pig pen near the house is, the smell; but the labor required to carry the slops of the house, etc., through a dirty barn-yard, would provide muck and other absorbent materials for rendering the pig pens free from all unpleasant odor, and furnish a large quantity of valuable manure into the bargain.

Pigs should have access to fresh water at all times, and the piggeries should be near a pump. If there is no well, a large cistern should be provided, and the rain-water from the buildings conveyed into it; and, in any case, the buildings must be furnished with gutters, to prevent the water running on to the manure, and washing out its soluble and most valuable plant-food.

Where stone is abundant, this is the cheapest and best material for the lower story of the pig pens. The floors may be laid with flags, and the joints grouted with water-lime and gravel, or the whole may be grouted with lime and gravel, taking care to provide good drainage. We should add, however, that a farmer of much experience, who built an expensive piggery, and flagged and grouted the bottom of the pens, says that his pigs did not thrive in them, and he subsequently put in plank floors. He thought the grouted floor was cold and damp. He is satisfied, at any rate, that pigs do much better on plank, than on stone, or grouted floors.

Some of our own pig pens have no other floor than *beaten earth;* and we are inclined to think that there is no material superior to it, and certainly none so cheap. The great point is, to have the ground high enough, so that the pens shall be always dry. If not so high that the liquid will not run rapidly away, draw on several loads of clay, and pound it down hard with a beater. Keep the pen well littered, and always clean; let the pigs have access to fresh earth, ashes, charcoal, etc., and they will not root up the floor.

In arranging the pig pen, special attention should be given to providing a ready means of cleaning out the manure, and supplying it with fresh bedding. A pig pen should be cleaned out every day, as regularly as we clean out our stables. If the pens are conveniently arranged for the purpose, it is but a few minutes work, and it will soon lead the pigs to form cleanly habits, and thus save bedding.

In pens for breeding sows, we have found it very convenient, in cold weather, to have a partition between the sleeping and feeding apartments, with a sliding door, that can be easily closed. It is desirable, when pigs are to be made ready for the butcher in eight or nine months, that the sow should farrow early in March; and it often happens that this interesting event occurs during a severe snow storm. With a warm sleeping apartment, and with a door that can be closed at night, or at any time after the sow has been fed, thousands of pigs that are now lost might be saved. This plan is particularly essential where the feeding apartment is partially or wholly uncovered. But even where both apartments are covered, it is better to have a partition that can be opened in warm weather, and closed during cold storms.

The only objection to this plan is, that the sow has not so much room, and there may be increased danger of her crushing the pigs against the sides of the pen. This objection, however, is more apparent than real, from the fact, that no matter how large the pen is, the sow is almost certain to make her bed near one of the sides. She almost invariably, in pigging, places her *back* against the rail or side of the pen, the object probably being to prevent the little pigs from getting on the wrong side of her, where they would, in cold weather, be likely to perish before they find the teats. Our breeding pens have a rail on the inside, about six inches from the sides of the pen and about one foot high, but the sows before pigging take special pains to fill the space with straw, and we are satisfied that if they did not, the little pigs, when born during a cold night, would often get on the backside of the sow and be chilled to death.

The accompanying plan of a piggery (fig. 29) is furnished us by Dr. M. Miles, Professor of Agriculture in the Michigan Agricultural College, who writes:

"It needs but little explanation, except in regard to the backside of the building. The lean-to is a shed, open above the pen partition, that separates it from the yard. This

Fig. 29.—PIGGERY AT THE MICHIGAN AGRICULTURAL COLLEGE. ELEVATION.

opening may be closed in winter, if desirable. The upright, or main building, is not boarded up below the roof of the lean-to. Figure 30 gives the ground plan. The curved, dotted lines, show the swing of the doors, and the straight, dotted lines, mark the position of the low partitions, enclosing the bed. The plan of arrangement can be carried out with a single pen, or it can be indefinitely extended for large establishments. The shed for protecting the manure can be readily cleaned out by a cart or wheel-barrow, running through the open doors, between the shed pens, while the swine are shut out in the yards, or in the front pens. I have not attempted to show the arrangement of the troughs, but simply mark their position. Swine can be easily changed from one pen to another, by shutting out others in the yard, or front pen. The upper story is for storing feed, or bedding, etc."

The writer's pig pens are of a very simple kind, put up by an ordinary farm hand, as a temporary arrangement,

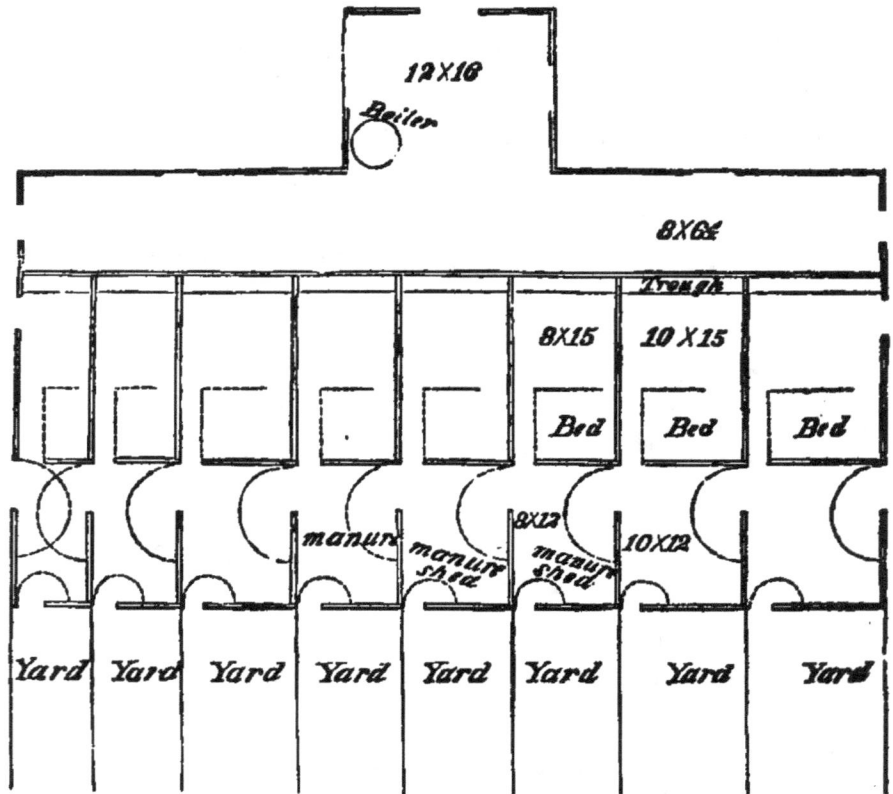

Fig. 30.—PIGGERY AT THE MICHIGAN AGRICULTURAL COLLEGE. GROUND PLAN.

but, as they are found to answer a very good purpose, in the absence of anything better, we give a description of them.

The old pig pen, which we found on the farm, was placed on one corner of the barn-yard, selected, apparently, because it was the lowest and wettest hole about the premises. The bottom was laid with plank, to keep the pigs out of the water. This was very well; but the moment the pigs stepped out of the pen, they plunged into a wet mass of manure and filth. They were obliged to wallow through this mud and manure every time they went to or from the pig pen. We have a weakness for hyacinths and roses, but found that the largest beds of them afforded no pleasure so long as there was such a pig pen in one corner of the garden.

Thanks to the invention of India rubber boots, it was possible to get about on the backside of the pig pen. We endured this two years, being determined not to fall into the common error of new-comers, of tearing down old buildings before we had determined where to erect new ones. At length, however, with axes, crowbars, a span of horses, and a log chain, we made short work with the old pig pen. Not a stick of it was left standing. The ground being cleared, the first thing was to dig an underdrain, 3 feet deep, underneath, and at the point where the surface-water settled; we covered the tiles with stones to the top of the land, so that the water from a heavy rain could pass off rapidly. We may add, that the soil underneath the old pig pen, for two feet deep, was found to be the blackest and richest of manure. With a plow and a dirt scraper, this was all removed, and ultimately drawn on to the land. This manure was certainly worth three times as much as the old pig pen.

The barn-yard was on a side-hill, the pig pen, as we have said, being on one of the lower corners. On the north side of the barn-yard there is a barn, with cow stables underneath, and a horse barn at the north-west corner. The pig pen was at the south-west corner. The first thing done, was to build a stone wall on the west side of the yard, 80 feet long, and 6 feet high, laid in mortar. The next thing was, to plow out the center of the barn-yard, and, with a dirt scraper, and a span of horses, make a basin 5 or 6 feet deep, with sloping sides. The dirt from this basin was emptied along the side of the stone wall, 15 or 16 feet wide, with a

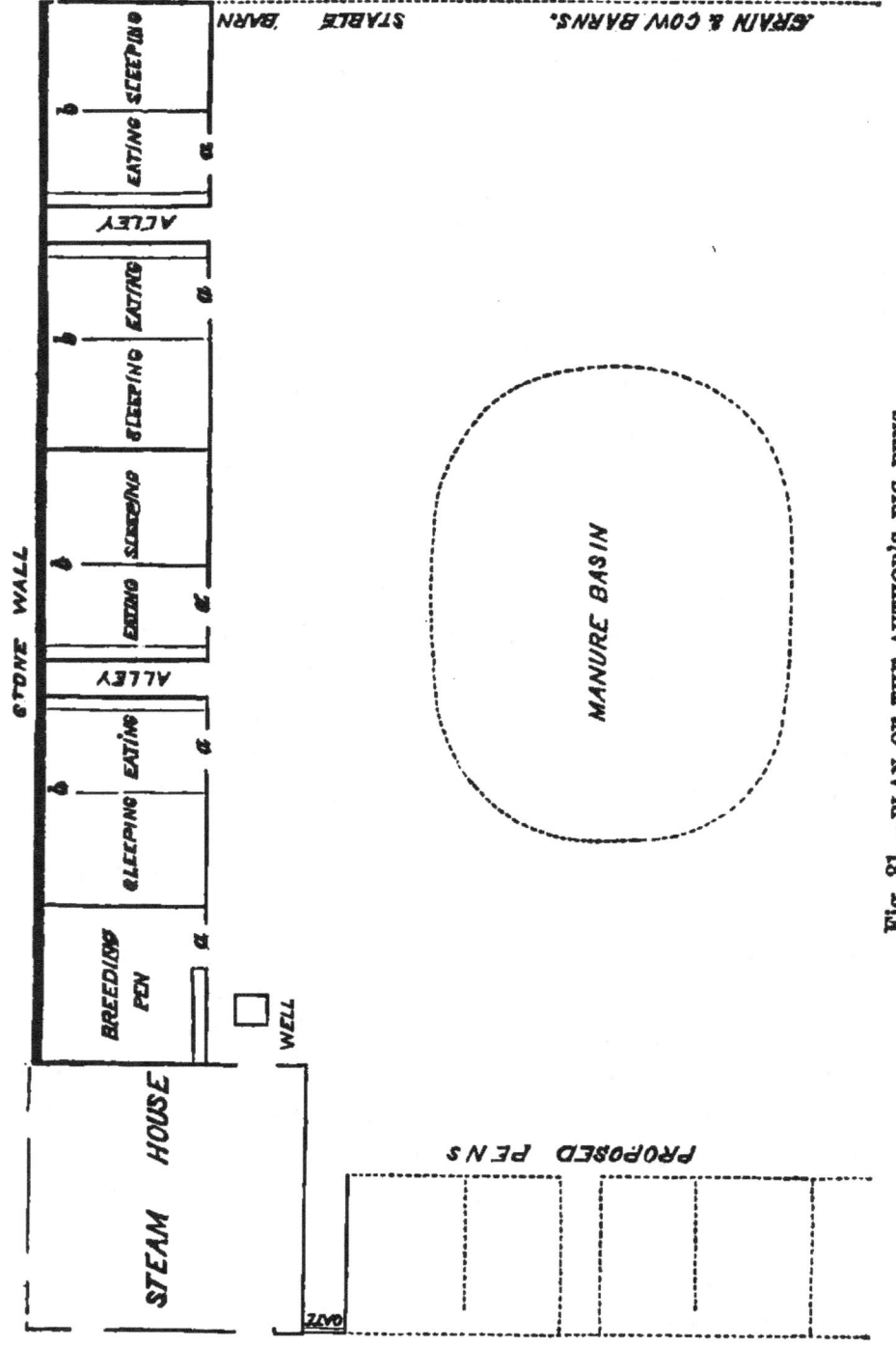

Fig. 81.—PLAN OF THE AUTHOR'S PIG PENS.

a, Doors opening from the pens into the yard; *b*, Doors between sleeping and eating apartments.

gentle slope from the wall, and the old hole, where the former pig pen stood, was raised in the same way. This gave us a dry foundation. As we have said, the wall was built 6 feet high, but, by the time we had scraped out the basin, and put the dirt on the side of the wall, we had raised the land 18 inches, or 2 feet. In other words, the land on the east side, towards the barn-yard, was nearly 2 feet higher than on the opposite side of the wall. The underdrain alluded to, runs along the side of the wall, on the west side, outside the barn-yard, and now, instead of needing India rubber boots, we can walk around in slippers.

On the top of the wall, a stick of timber was placed, and we proceeded to put up a common shed, with roof boards, 14 feet long, and battened in the ordinary way. Of course the roof slopes towards the wall, so as to carry the water outside of the barn-yard, where it soaks through the soil to the underdrain. This shed is divided off into pig pens, as shown in the diagram, figure 31.

The pens are 12 feet deep, and 16 feet wide. (It would have been better to have had the roof boards 16 feet long instead of 14 feet, as it would have added very little to the expense, and would have given us pens 14 feet deep.) Between each two pens is an alley, 3 feet wide, boarded up on each side about 3 feet high. The pig trough is placed along-side these partitions, and the food is poured into it from the alley.

Each pen is divided off into two parts—one for sleeping, and the other for feeding. The sleeping apartment is boarded up tight, with a sliding door, against the wall. One of the boards that forms the partition between the feeding and sleeping apartment is hung on hinges, so that it can be opened or shut, according to the weather. It is fastened by a common wooden button. One of the boards which form the outside of the pen is hung in the same way. This is very important, as it enables us to give an abundance of fresh air in warm weather, and we can close up the pen tight during a storm. It is also convenient in cleaning out the pen and putting in fresh bedding.

We do not recommend these pens to any one who can afford to build better ones. Their chief merit consists in their cheapness. They can be easily cleaned out, and supplied with fresh litter. Our pigs, when old enough, are allowed to run out every day, into the barn-yard, in winter, and the pasture in summer; and we find this arrangement convenient for letting them in and out of the pens, as each pen opens directly into the barn-yard. If well-bred, and *properly treated*, the pigs will go to their own pens as readily as cows or horses will go to their own stalls. This may be doubted by those who ill-treat their pigs—or, in other words, by those who treat their pigs in the common way. But it is, nevertheless, a fact, that there is no more docile or tractable animal on a farm than a well-bred pig. There is a good deal of human nature about him. He can be led where he cannot be driven. A cross-grained man will soon spoil a lot of well-bred pigs. They know the tones of his voice, and it is amusing to see what tricks they will play him. We have seen such a man trying to get the pigs into their respective pens, and it would seem as though he had brought with him a legion of imps, and that seven of them had entered into each pig. No sow would go with her own pigs, and no pigs would go with their own mother; the store pigs would go into the fattening pen, and the fattening pigs would go where the stores were wanted. Should he get mad, and use a stick, some active porker would lead him in many a chase around the barn-yard; and when one was tired, another pig, with brotherly affection, would take up the quarrel, and the old sows would stand by enjoying the fun. Let no such man have charge of any domestic animals. He is a born hewer of wood, and drawer of water, and should be sent to dig canals, or do night-work for the poudrette manufacturers.

PIGGERIES AND PIG PENS.

At their regular feeding time, we can take twenty or thirty of our own pigs, and separate them into their respective pens in a few minutes. They *inherit* a quiet disposition, and we would dismiss on the spot, any hired man who should kick one of them, or strike him with a stick, and we cannot bear to hear an angry word spoken near the pens.

The alleys between the pens we find convenient for storing away a small quantity of straw, a little of which can be used every day, to replace that removed in cleaning the pens. By making a small hole in the side of the pen, little sucking pigs can come through, and eat a little milk or crushed oats out of a small trough, placed in the alley where the sow cannot get at it.

We have some pens that have no partition between the sleeping and feeding apartment. They are not as warm as the others, but having abundance of straw, they answer very well for store or fattening pigs or for a breeding sow in mild weather. On the whole, however, it is better to have the sleeping apartment separate, the pigs being warm, and not so liable to be disturbed.

For a breeding sow, the sleeping apartment is 10 × 12 ft., and the feeding apartment 6 × 12 ft. Such a pen can be used also for six or eight store pigs, or for three or four fattening pigs.

We have smaller pens, 12 × 12 ft., either undivided or divided into a sleeping apartment 7 × 12 ft., and a feeding apartment 5 × 12 ft. Such a pen, if divided, answers very well for a litter of young pigs, after weaning, or for fattening two or three pigs, and we have used them for a small sow to farrow in.

The most serious objection to this shed-made pig pen is, that the roof boards must be put on with great care, and well battened, or it will leak. They should, also, be well saturated with petroleum, to keep them from shrinking and warping.

7*

Paschal Morris, of Philadelphia, an extensive breeder of Chester Whites, describes his plan of a piggery as follows:

"The plan of the piggery, delineated in the accompanying engraving, (fig. 32) is susceptible of reduction or extension, for a larger or smaller number of pigs, and is intended to supersede the not only useless, but objectionable, as well

Fig. 32.—PASCHAL MORRIS' PIGGERY.—ELEVATION.

as expensive, mode of constructing large buildings under one roof, where confined and impure air, as well as the difficulty of keeping clean, interfere greatly with both health and thrift. Twenty-five or thirty breeding sows, farrowing at different periods of the year, can be accommodated under this system of separate pens, by bringing them successively within the enclosure, or an equal number of hogs can be fattened, without crowding or interference with each other.

"The entrance, as seen in the engraving, is on the north side of the building, which fronts the south, as does also each separate pen. The main building is 32 feet long, by 12 feet wide, with an entrance gate, at each lower corner, to the yard of two first divisions. The entry, or room, in the center, is 8 feet wide, allowing space for slop barrel, feed chest, charcoal barrel, (almost as indispensable as feed chest,) hatchway, for access to root cellar, underneath the whole building, and also passage-way to second story. This latter is used for storing corn in winter, and curing

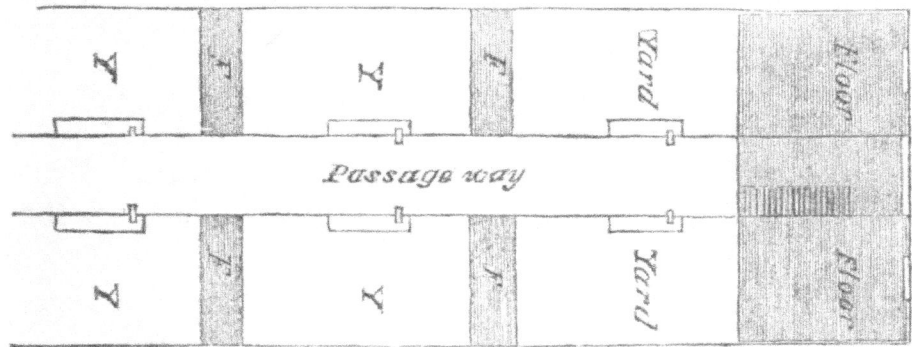

Fig. 33.—PASCHAL MORRIS' PIGGERY.—GROUND PLAN.

some varieties of seeds in summer. A wooden spout, with sliding valve, conveys feed to the chest below. The grain is hoisted to the second floor by a pulley and tackle on the outside, as observed in engraving.

"The perspective of main building allows a partial view of platforms, surmounted by a board roof, and divisions in the rear. The ground plan, fig. 33, allows six of these on either side of the passage-way. The first two pens, to the right and left of the door, are 12 × 12 feet each, and attached to them are 25 feet in length of yard, by 15 feet wide.

"All the yards are extended 3 feet wider than the building, which admits of the two entrance gates at the corners.

"Another division then commences, consisting of a raised platform, 6 to 8 feet wide, and extending the same width as the first pen, with a board roof over it, and also boarded up on the back, which answers the purpose of a division fence, to separate from the pen behind. Twenty-five feet of yard are also attached to this, and the same arrangement is continued to all the six divisions.

"We have found this board roof and wooden floor, on the north side of each pen, and fronting the south, to be ample protection in cold, wet, or stormy weather. The floor is kept perfectly clean, and even the feeding trough is not on it, on account of more or less of wet and dirt, always contiguous to the trough, which freezes in winter, and becomes slippery.

"Each yard is used for the deposit of refuse vegetables and weeds, litter, etc., thrown in from time to time, to be consumed or converted into manure. This is conveniently loaded into a cart, passing along on the outside of each range of pens.

"The passage-way between each range of pens gives convenient access to the feeder for all the divisions. A door also communicates from one division to the other, to make changes when necessary; and also a door, or gate, from each pen to the outside, so that one or more can be removed, and others introduced, without any confusion or interference from any of the other pens. The two pens under the main roof of the building, being more sheltered, are reserved for sows who may happen to farrow very early in the season, or in extreme cold weather, which is always avoided, if practicable.

"For several reasons, the boiler for cooking food is in a rough shed, adjacent to the piggery, and entirely outside of it. There is no reason why this should be necessarily a part of the piggery.

"The above plan is not offered as embracing much that

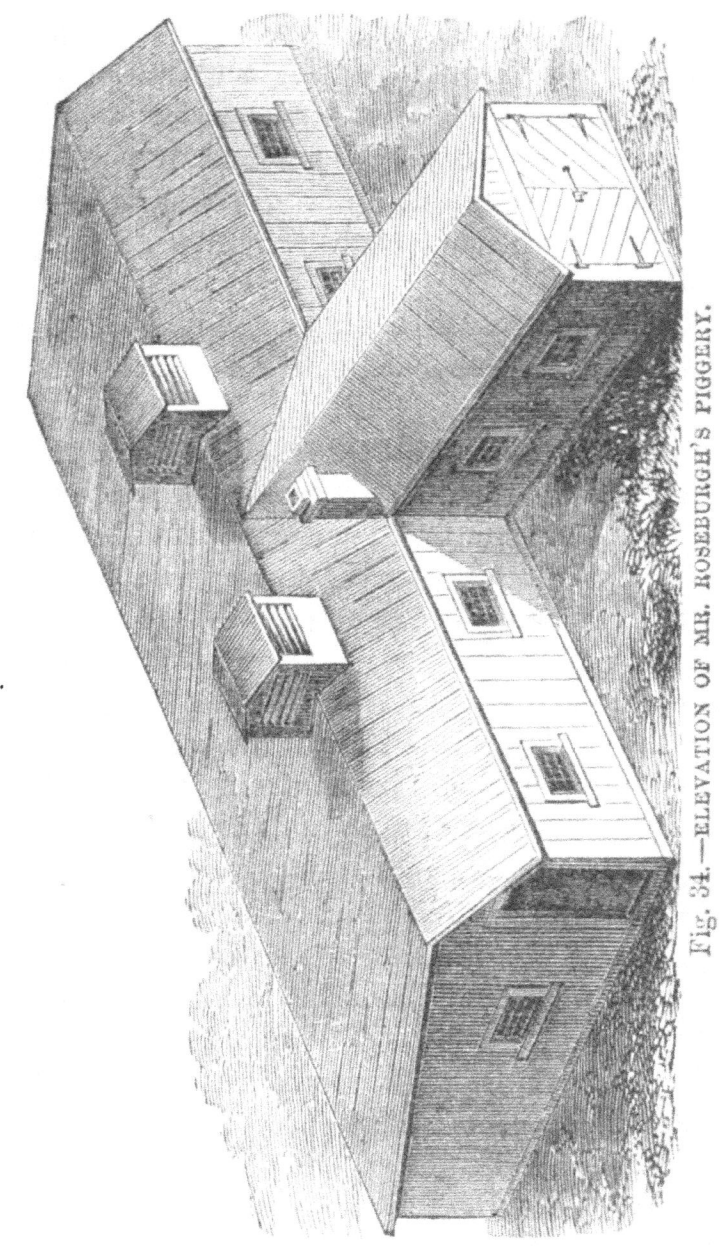

Fig. 34.—ELEVATION OF MR. ROSEBURGH'S PIGGERY.

is novel in arrangement, but as one that combines many advantages—

"1st. Complete separation, as well as easy communication between each pen, as well as to outside from each.

"2d. Avoiding close and confined air, and admitting of extension or alteration for a large or small number of pigs.

"3d. Facilities for keeping clean and receiving refuse vegetables and weeds, etc., for conversion into manure, and also for loading from each pen into a cart, passing along outside.

"4th. Cheapness. With the exception of the main building, all the rest can easily be erected by an intelligent farm hand."

The illustrations (figs. 34, 35 and 36) were engraved for the *American Agriculturist*, from plans forwarded by Mr. Roseburgh, of Amboy, Ill. They were designed and constructed for use on his own premises, and have, therefore, the merit of being the production of a practical man.

Fig. 34 represents the elevation. The main building is 22 by 50 feet, and the wing 12 by 16 feet. It is supplied with light and air by windows in front, ventilators on the roof, and by hanging doors or shutters in the upper part of the siding, at the rear of each stall or apartment—these last are not shown in the engraving.

Fig. 35 shows the ground plan. The main building has a hall, *H*, 6 feet wide, running the entire length. This is for convenience of feeding, and for hanging dressed hogs at the time of slaughtering. The remainder of the space is divided by partitions into apartments, *A*,*B*, for the feeding and sleeping accommodation of the porkers; these are each 8x16 feet. The rear division of each apartment, *B*, *B*, is intended for the manure yard: Each apartment has a door, *D*, *D*, to facilitate the removal of manure, and also to allow ingress to the swine when introduced to the pen. The floors of each two adjoining divisions are inclined toward each other, so that the liquid excrements and other filth may flow to the side where the opening to the back apartment is situated. Two troughs, *S, T*, are placed in each feeding room. That in the front, *S*, is for food, and *T*, for clear water, a full supply of which is always allowed. This is an important item, generally overlooked; much of the food of swine induces thirst, and the free use of water is favorable to the deposition of fat.

An excellent arrangement (shown in fig. 36,) is adopted

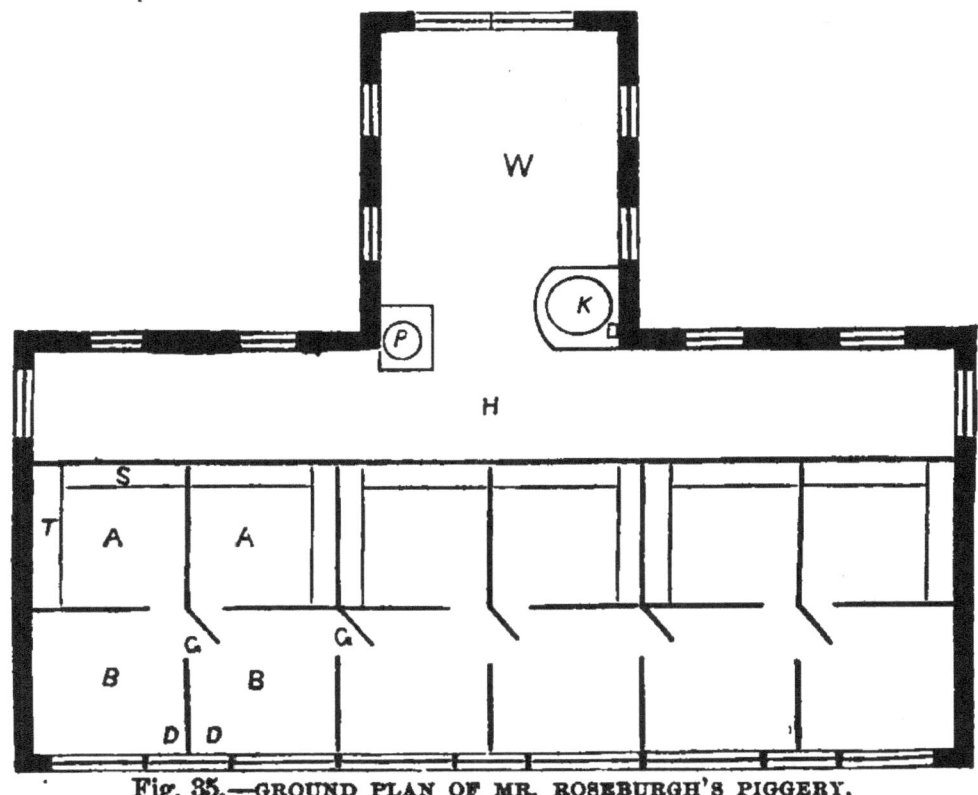

Fig. 35.—GROUND PLAN OF MR. ROSEBURGH'S PIGGERY.

to facilitate the cleaning of the troughs, and the transferring of the hogs to the main hall at slaughtering. The front partition of each apartment, *F*, (fig. 36,) is made separate, and contrived so as to be swung back, and fastened over the inside of the trough, *T*, at feeding time, or when cleaning the trough. It may also be lifted as high as the top of the side partition, *H*, when it is desired to take the hogs to the dressing table. Triangular pieces, *E, E,* are spiked to each front partition, and swing with it, forming stalls to prevent their crowding while feeding. These are supported, when the apartment is closed, by notches in the inner edge of the trough, made to receive them.

The wing, *W,* (fig. 35) is 12 by 16 feet. This answers for a slaughtering room. In one corner, adjoining the main hall, is a well and pump, *P,* from which, by means of a

Fig. 36.—VIEW OF FRONT PARTITION.

hose, water is conveyed to the troughs. At the opposite corner, *K,* is a large iron kettle, set in an arch, for cooking food, and for scalding the slaughtered swine. We would suggest that, in many localities, it would be a desirable addition to have this wing built two stories high, the upper part to be used for storing grain for the hogs, and also that a cellar be made underneath for receiving roots.

We give from the *American Agriculturist* illustrations taken from the working drawings of a pig-house which has recently been built at Ogden Farm (Newport, R. I.). It is submitted to those of our readers who may contemplate improvements of this sort. The building is 14 × 32 feet, and cost (built of rough pine battened, with cedar shingles on the roof) only $425, including the excavation of the manure pits, and the boarding up of their sides.

"Fig. 37, is the ground plan. There are four pens 8 × 10, two 6 × 10, and two 6 × 12. The troughs all open into the center area, and are opened by swing posts, which expose them to the attendant for cleansing or filling, or to the swine for feeding, as may be desired. The two large bins at the sides of the entrance door are filled with dry earth, with which the pigs are treated to the luxury of the earth-

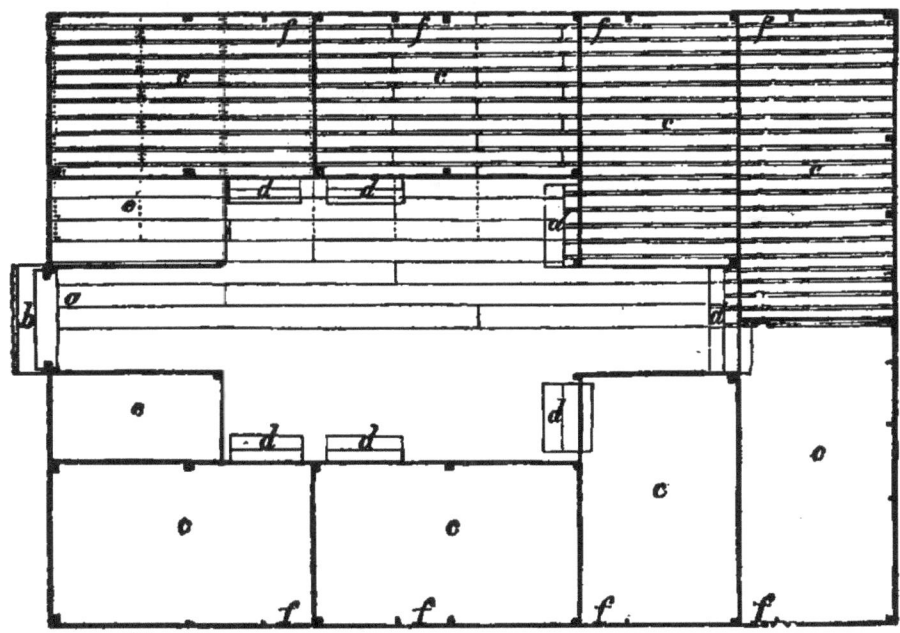

Fig. 37.—GROUND PLAN OF OGDEN FARM PIGGERY.

closet—to the great improvement of the air of the building, and of the manure. The floors of the pens are made of 2-inch planks, 6 inches wide, laid with 1-inch openings between them, which secures the immediate passage of the urine to the pits below, and the gradual working through of the dry manure, mixed with earth. In the center of the open floor stands a Prindle steamer, whose 7-inch smoke-pipe discharges into the middle of a 12-inch galvanized iron ventilator, whereby efficient ventilation is secured. The food is cooked in pork-barrels, which may be moved about at pleasure; the flexible steam hose, with an iron nozzle, conveying the steam to the bottom of the barrel. Figure 38 is a cross section, showing the manure

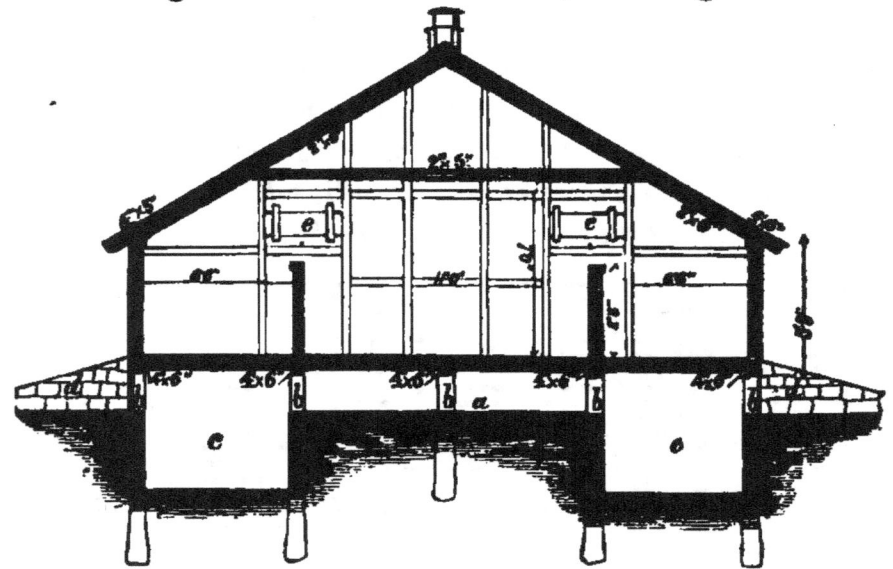

Fig. 38.—CROSS-SECTION OF OGDEN FARM PIGGERY.

pits, pens, etc. More than fifteen cords of manure can be stored in the pits, which are to be emptied through shuttered windows. Figure 39 is the front elevation of the

Fig. 39.—FRONT ELEVATION OF OGDEN FARM PIGGERY.

building, which is to have small yards at the sides, communicating with the pens by slopes from the outer doors. This house will accommodate from thirty to forty shoats, or a corresponding number of breeding animals."

Mr. Geo. Mangles, a very extensive breeder and feeder of pigs in Yorkshire, England, has constructed a cheap and simple shed for fattening pigs, engravings of which, taken from Mr. Sidney's edition of Youatt on the Pig, we annex. Mr. Mangles' description is as follows:

"For feeding pigs the best arrangement is a covered shed (shown in figure 40), kept dark, with partitions to hold three pigs in each division, as feeding-pigs do not require much exercise. If the pigs be fed regu-

Fig. 40.—MR. MANGLES' COVERED SHED FOR FATTENING PIGS.

larly, and a little fresh bedding spread every day, the animals sleep and thrive very fast. The improvement they make in a warm, covered shed, with plenty of fresh air, is astonishing. A feeding-pig cannot be too warm, if he has plenty of fresh air.

"I have had pigs fatten very fast upon latticed boards, with pits underneath for the droppings. The boards should be swept occasionally, and sawdust sprinkled over

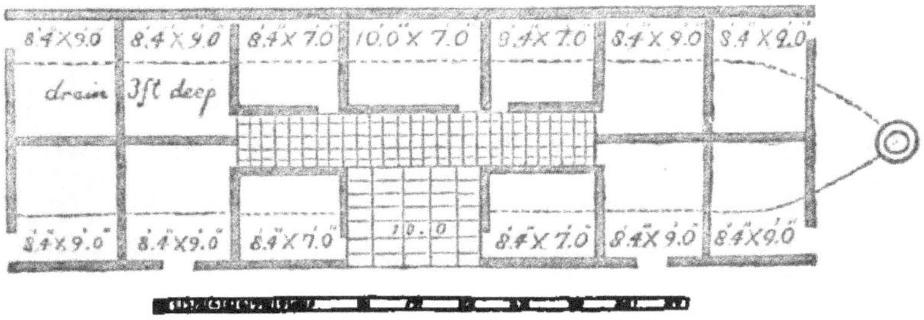

Fig. 41.—MR. MANGLES' SHED.—GROUND PLAN.

them and swept through. This plan will only do for feeding-pigs (not for pigs for sale, breeding, or exhibition), as their houghs swell very much; but young pigs always do better on boards than on stone floors.

"The covered pig-shed (fig. 41), of which a plan accompanies this description, will hold about sixty pigs; the roof is of light spars, covered with felt, but thin boards would be better and cheaper in the end. The pigs thrive in an extraordinary manner in this shed, which is divided into nineteen pens, of different sizes, some of which I find useful at lambing time to put ewes and lambs in at night."

DESCRIPTION OF ISOMETRICAL PLAN OF PIG-SHEDS, (Fig. 42,) SHOWING THE INTERNAL ARRANGEMENTS.

"Length of shed, 60 feet; breadth, 18 feet, inside; height of walls (*of brick*), 6 feet; height of pens inside, 3 feet, 6 inches; thirty-three posts, 9 feet long, and 3 inches square out of ground; five posts, 5 feet long, by 3 inches; two strong posts for doors, 6 inches square.

Pens.

4 rails, 13 feet long, 3 inches by 1½ inches.
8 " 9 " " "
14 " 8 feet, 4 in., " "
8 " 7 feet " "
4 rails 6 " " "
4 " 5 " " "

600 poles, 3 feet 6 in. long, 3 in. by 1 inch.
90 feet boards, 11 in. by 1 inch.
150 boards for doors, 11 in. by 1 inch.

"*Wood-work for Roof.*—Three boards for the center, to nail rafters to, 20 feet long, 9 inches deep, and 1 inch thick; sixteen rafters, 13 feet long, 3 inches by 2; 58 rafters, 13 feet long, 3 inches by 1¼; 120 feet of rails, 3 inches by 1½, to lie on wall, to nail rafters to; eight rails, 20 feet long, 3 inches by 1½; ten lengths of felting, 60 feet long; 1,660 feet boarding, required 11 inches broad.

"There are air-holes in the brick walls to every pen, on one side; on the side where the folding doors are set, there are four air-holes, and two holes for throwing the manure out. One end of the shed is boarded half way

PIGGERIES AND PIG PENS.

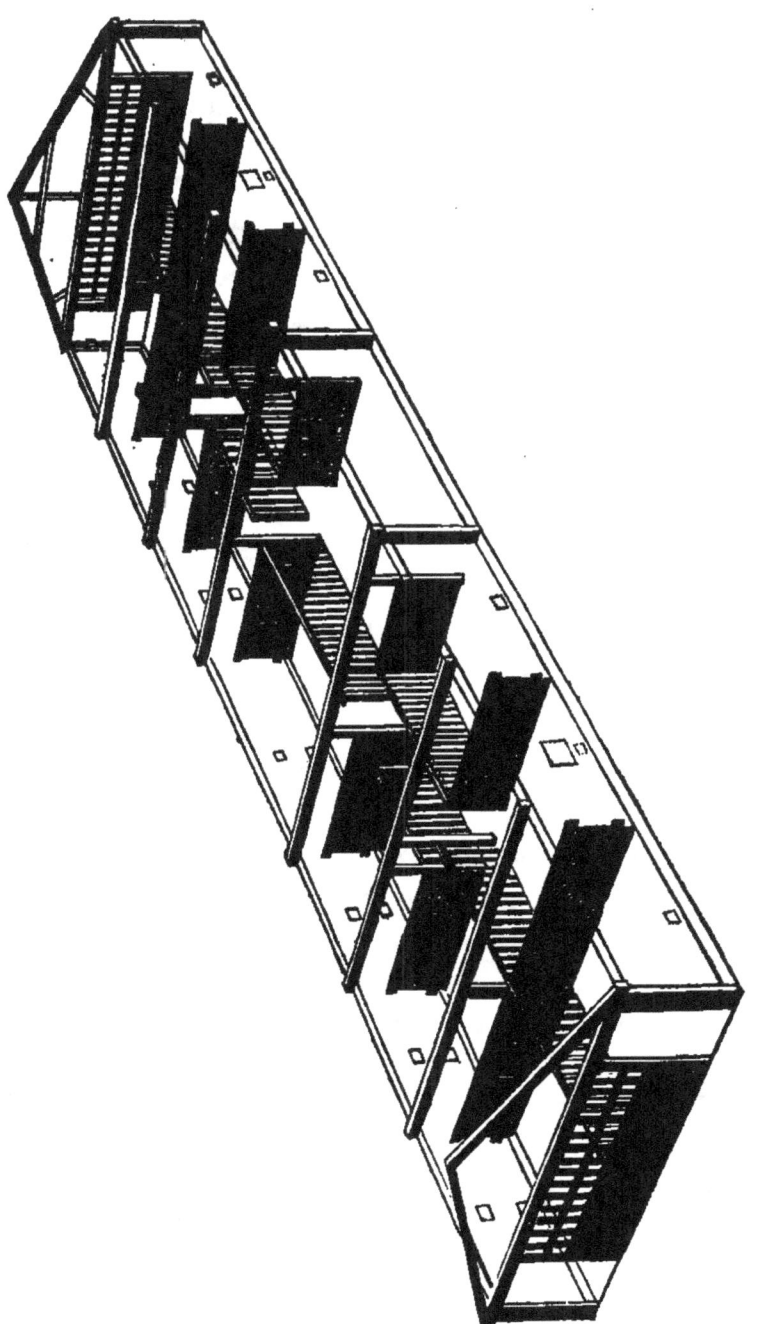

Fig. 42.—ISOMETRICAL VIEW OF MR. MANGLES' PIGGERY.

up, the rest of it up to the point of the gables of open palings; the other end is boarded, and a large space is filled with Venetian blinds, or *louvres.*

"The floor of the pens is of beaten soil; a drain, 3 feet deep, filled with stones, leads to the liquid manure pit. The passage is laid with bricks, and the entrance is flagged, and a cart can be backed up to take the manure when the pig pens.or pits are cleaned out. I generally let the pits get full of manure, and contrive to empty

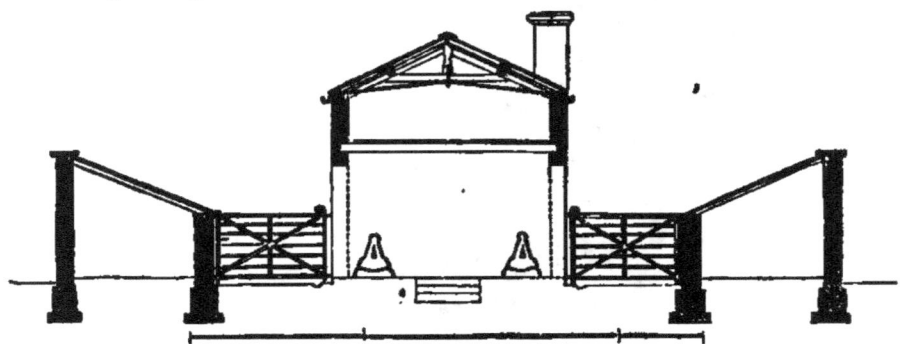

Fig. 43.—SECTION OF COVERED FOOD HOUSE OF TATTENHALL PIGGERY.

them against the turnip season. They are soon emptied; it takes one hand more than the ordinary force for filling manure.

"I whitewash the walls and partitions every year, and the man keeps the passage swept and covered with sawdust. My troughs are iron, with many divisions, and filled by hand from the passage. Each pit will hold five or six porkers, or three bacon pigs."

One of the most elaborate piggeries in England is that at Tattenhall Hall, in Cheshire, forming a part of the model farm buildings on a dairy farm of 330 acres, in the occupation of Mr. George Jackson. The pig sheds are each six feet high, and the feeding troughs, and the passage alongside them, are under cover.

Figure 43 gives a section through the food-house, and figure 44 a ground plan of the arrangement.

"The floors of the pig-yards and the pig-sheds are of

strong sandstone flags. The two near sheds are provided with doors, to keep them warm in cold weather, and with iron doors, fifteen inches square, set in the outer wall, for ventilation in hot weather. A joist is set on three sides, one foot from the wall, and one foot from the floor, to

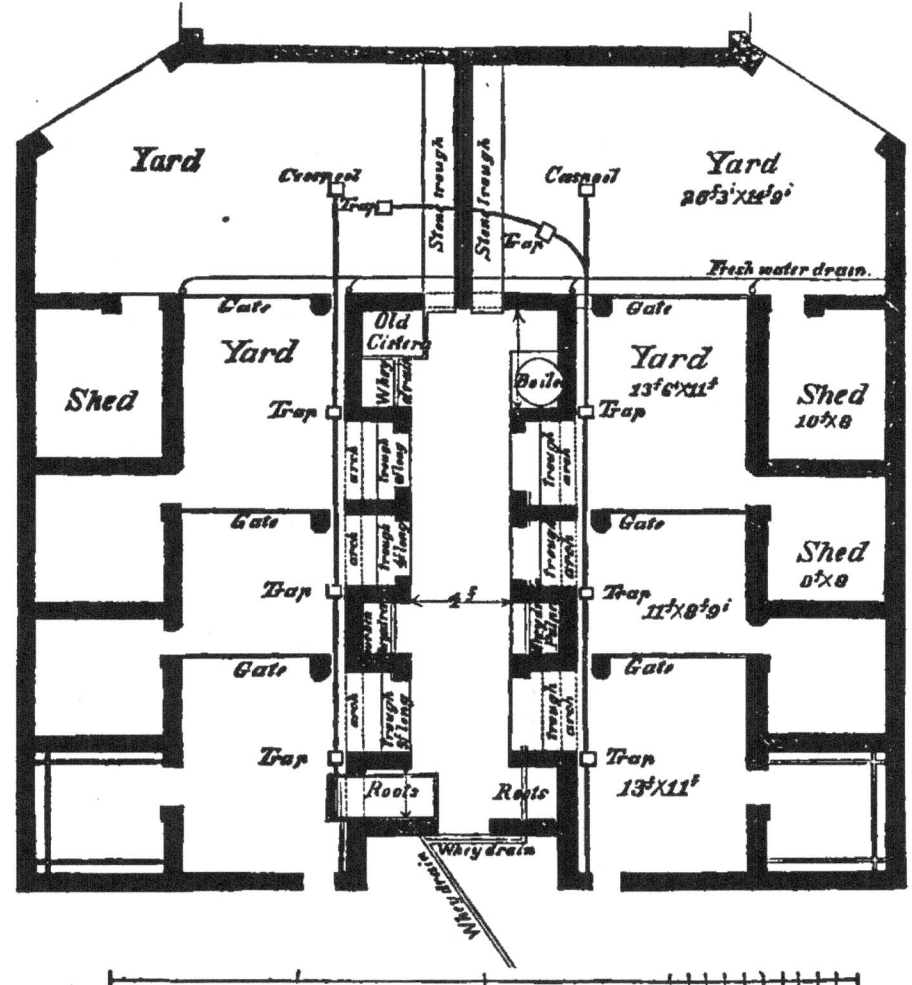

Fig. 44.—GROUND PLAN OF TATTENHALL PIGGERY.

prevent mothers from overlaying their young. The 'outlets,' or yards, are too small; but we were cramped for space. The drains to all the liquid manure-tanks are trapped.

"'Whey,' says Mr. Jackson, 'forms the staple food of my pigs, the fatting ones getting a portion of Indian corn-meal and barley-meal, with, occasionally, in winter, roots.'

"It will be seen that the food-house is the receptacle of these kinds of food. The Windsor troughs, with swing doors, push back, and shut out the pigs while the solid food is put into the troughs, and one key locks up the whole. The whey is laid on to all the troughs from four large whey-cisterns in the buttery, and one hundred pigs are, all summer, daily fed with as many gallons of whey per meal, *in one minute, by simply lifting a valve*. By this plan is *pig-feeding made easy*, and they get properly, instead of laboriously and irregularly, fed. The iron gates are provided for enabling to cleanse and straw the sties. The rain-water goes off by a drain, and the liquid manure passes to the 'tank,' from which it is drawn by drain, at pleasure, into a liquid manure cart, in the middle of a ten-acre meadow. The fowls are over the food-house, the floors of which are flags, but are equally adapted for boards."

These plans are given merely for the purpose of furnishing useful hints. Each farmer must determine for himself what kind of pig pens are best suited to his wants—to his location, system of feeding, etc. But whatever plan he may adopt, he should recollect that dryness, warmth, and good ventilation, are absolutely essential to the best success in pig feeding.

There is one point in Mr. Mangles' plan that is worthy of consideration, and that is, the "beaten soil" for the floors of the pens, and the stone drain, three feet deep, under the pens, to carry the drainage to the liquid manure pit. Where such thorough drainage is provided, there can be no doubt that earth floors, beaten hard, answer a good purpose, and save much expense. When the floors are made of plank, they soon get worn in holes, and the liquid soaks through the joints; and if not ultimately lost, we loose the use of it for several years, or until the pen needs a new floor, and the soil underneath is thrown out and replaced with fresh earth. With beaten clay floors, very little liquid will soak into the earth, and if it does, the plant-food which it contains would be absorbed near the surface, and, by scraping the floors, it would all find its way to the manure heap.

CHAPTER XVII.

SWILL BARRELS, PIG TROUGHS, ETC.

In some convenient place, near the pig pens, there should be a receptacle for the wash from the house, milk, whey, waste vegetables, and other refuse. This is often nothing more than an old pork or cider barrel. It is difficult to conceive of anything more inconvenient. It is too high, and too circumscribed. A far more convenient and inexpensive arrangement is to make a tub out of two-inch pine planks—say six feet long, two feet and a half wide, and two feet, or two and a half or three feet high—according to the number of pigs kept. Or, what is better still, make such a tub out of plank twelve feet long, and have a partition in the middle. In this way you have two tubs in one. The food for the store pigs can be kept in one, and that for the fattening pigs in the other. In our own case, we find it desirable to have two such tubs, each twelve feet long, and divided in the middle. Such tubs are often made flaring, being wider at the top than at the bottom. We do not think there is any material advantage in this, and it requires more skill to make the grooves fit true, and it is not so easy to furnish them with a tight-fitting cover. The latter is very desirable. It should be put on with hinges, and made of planed and matched inch boards, and divided in the center of the tub, so that one part may be closed while the other is open, if desired.

At the house, a barrel should be placed in some convenient place, for the reception of all dish-water and refuse. If this barrel is set on wheels, as shown in the engraving, fig. 45, copied from the *American Agriculturist*, it can be easily conveyed to the pig pens, and emptied into one of the tubs above described. It should then be mixed

Fig. 45.—PORTABLE SWILL BARREL.

with a little meal, and allowed to remain until the particles of meal become quite soft. It is then much more easily digested. If a slight fermentation takes place, by which the starch of the meal is converted into sugar, and a little of it into alcohol, the pigs appear to relish it all the better. A small amount of meal fed to store pigs in this manner, in summer, enables us to obtain

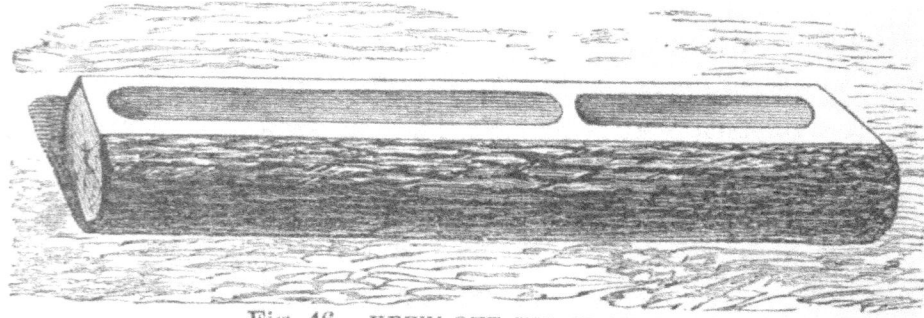

Fig. 46.—HEWN-OUT PIG TROUGH.

much more benefit from the milk, whey, and house wash than when fed alone.

Every pig pen should be provided with two troughs—one for food, and the other for water.

When wood is abundant, the commonest, and perhaps

the cheapest pig trough, is made by taking a log about fifteen inches in diameter, and, with an axe and adze, hewing out the inside. For out-door feeding, they are the most convenient troughs we are acquainted with, as they are not easily upset.

When used for pigs confined to pens, the log should be hewn out in two divisions, one for food, and the other for water, as shown in fig. 46. A twelve-foot log will give about six feet of trough for food, and two and a half to three feet for water.

A better and equally simple pig trough is made from two-inch pine or hemlock planks. The planks should be from nine to fifteen inches wide, according to the size of

Fig. 47.—PLANK PIG TROUGH.

the pigs, and the number in a pen. The planks are nailed firmly together at right angles, with twenty-penny nails, put nine inches apart. There should be either two troughs for each pen, or the one trough should be divided into two compartments, one for water, and the other for food. The ends of the plank must be sawed off square and true, and a piece of plank nailed at each end, sufficiently tight to hold water. Such a trough is much more likely to leak at the ends than at the bottom, and great care should be taken to saw them off square, and nail them on tight. When both planks are the same width, the plank that is to be against the side of the pen, and farthest from the pigs should, in nailing, be placed on the other. This will make that side of the trough two inches higher than the one next the pigs, and they will be less likely to waste the food. The end pieces should project about four inches beyond the edge of the trough, as shown in fig. 47. This allows it to stand so firmly that the pigs will not be likely to upset it.

Before being used, the troughs and the swill tub should be *thoroughly saturated* with petroleum. This will not only preserve the wood, but do much to prevent it from warping, and the pigs will not be so likely to gnaw holes in the troughs.

The *American Agriculturist* gives the following plans of pig troughs which allow the food to be distributed along the trough from the outside:

"The pens (fig. 48), being made of horizontal boards,

Fig. 48.—A CONVENIENT PIG TROUGH.

nailed to posts about 6 feet apart, the troughs are accurately fitted between two posts, so as to project a little outside the boarding, and the board above the trough is nailed on a little above it; so that, when the edge is chamfered off a little, any thing may be easily poured into it throughout its whole length. This arrangement admits of putting partitions, nailed to the pen above the trough, and to the floor, dividing the trough into narrow sections, so that each pig shall get only his share. The only objection to this form of trough is, that it must be cleaned out from inside the pen.

"A modification of this arrangement may be made, the trough coming flush with the outside boarding, and the board above it being simply taken off and nailed on the inside of the posts, and stayed by a piece nailed perpendicularly, so as to stiffen and prevent its springing.

"In figure 49 we show an old plan which, after all, is one of the very best contrivances for hog troughs. The

Fig. 49.—SWINGING DOOR PIG TROUGH.

trough is set projecting somewhat outside the pen, and placed as in the other pen, filling all the space between two posts. Over the trough is hung a swinging door or lid, some 3 feet wide, and as long as the trough. A wooden bolt is placed upon this lid, so that when it is swung back and bolted, the hogs are shut out completely from the trough; and when it is swung out or forward and bolted, they have access to it again. This style of trough is very easily cleaned out. The lid may have iron rods, beat into a V-shape, and having flattened ends, turned in opposite directions, screwed upon it, and so placed that they will entirely separate the hogs—when feeding. This contrivance is shown in fig. 50. Some arrangement of this kind will be found as great a conven-

Fig. 50.—SWING DOOR WITH FENDERS.

ience as it is an economy. The patented hog troughs are usually expensive, and no better, if so good. For our own use, we greatly prefer these simple fixtures, which may be easily made, renewed, or repaired, as occasion may require, with the common tools which every farmer

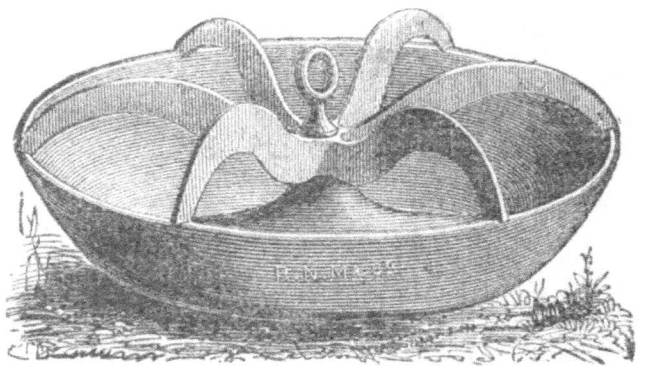

Fig. 51.—CAST-IRON PIG TROUGH.

should have and know how to use. Cast-iron pig troughs, of different patterns, are sold at the agricultural implement stores. One of them is shown in fig. 51; the weight of the one figured is one hundred and ten pounds.

CHAPTER XVIII.

MANAGEMENT OF PIGS.

The object of keeping pigs differs in different places and circumstances. The dairy farmer keeps pigs principally for the purpose of turning his whey and skimmed milk to good account. The grain-growing farmers, in the older settled parts of the country, keep pigs to consume the slops of the house, and to pick up scattered grain around the barns and on the stubbles, and to consume, and turn into pork, small potatoes, and many other articles that would otherwise be wasted. At the West, where corn is cheap, and the expense of sending it to market very great, pigs are kept for the purpose of "packing fourteen bushels of corn into a three-bushel barrel." In the vicinity of the Atlantic cities, pigs are kept, or might be kept, for the purpose of manufacturing out of purchased food, nice, fresh pork, and rich, valuable manure. And, indeed, in all sections where pigs are kept, the value of the manure should be taken into consideration.

PIGS ON DAIRY FARMS.

There is no other food on which young pigs thrive so well as on skimmed milk and Indian meal. Pigs are also very fond of whey, and do well on it *provided* they have a liberal allowance of pea-meal and Indian meal fed with it. To keep pigs on whey alone is a great waste of food and time. On skimmed milk, and the run of a clover pasture, a well-bred, young pig, will grow rapidly; but even in this case a little corn-meal could be fed with very decided economy and advantage. The oil and starch of the corn restore to the skimmed milk the fat-forming material which has been removed in the butter, and, in effect, convert it into new milk again. But it is very desirable that the meal should be cooked by pouring upon it boiling water, and stirring it carefully until it is made into "pudding." In the dairy there is usually much hot water thrown away, which might be used for this purpose, without cost, and with little labor.

Since the introduction of cheese factories, dairy farmers cannot keep as many pigs through the summer as formerly, but early in the spring, before the factories commence operations, the milk is used at home; and it is well to have some litters of young pigs, which can be sold to good advantage soon after weaning. The sows can be summered on grass and on the slops of the house, and another litter would be obtained in the fall. When cows are well wintered, and fed on more or less grain or oilcake, then fall pigs can be kept through the winter in good condition at very slight expense, and they will be valuable to sell to the factories or other feeders the next summer. Usually, this system will pay better than attempting to fatten them at home.

PIGS ON GRAIN FARMS.

On farms where much grain is grown, and only a few cows are kept, it is usually not profitable to keep a large stock of pigs. The common mistake made, however, is not in keeping too many, but in not feeding them liberally. As a rule, the pigs are kept on short allowance until they are shut up to fatten, after the corn is ripe, although there can be no doubt that a bushel of corn, fed to pigs while on clover during the summer, will produce double or treble as much pork as a bushel of new corn fed in cool weather, in the autumn, when the pigs have nothing but corn. A few fall pigs can be kept in the yards during the winter to good advantage, especially if the cattle are fed grain. But it is a great mistake to stint young pigs through the winter, although it must be confessed that it is a very common one. The sows, and any spring pigs that may be wintered over, will pick up the lion's share of the scattered grain and other food in the yards; and while it is often inconvenient to separate the young pigs from the older ones, yet it is not a difficult matter to make a hole in one of the sides of the pens that will admit the young pigs through, and exclude the large ones, and in this way the young pigs can be fed more and better food. This is a very important point. The young pigs should be kept growing rapidly through the winter and spring months. They should be in a condition that most farmers would pronounce "too fat." Young, wellbred pigs, so wintered, can be summered in a clover pasture at comparatively little cost, and it is astonishing how fast they will grow. We have kept a lot of grade Essex fall pigs during the summer on a rich clover pasture near the barn-yard, and the slop from the house, without any grain, that were sold at an extra price on the first of October, to "top-off" a car load of fat pigs sent to the New York market. And the whole secret of the matter, if secret it is, was in feeding the young pigs liberally through the winter.

Few things would pay a grain growing farmer better than to raise peas for his pigs. No matter how "buggy" the peas may be, the bugs or beetles remain in the peas until about the first of November; and when the peas are fed out before this time, the pigs will eat peas and bugs together, and there will be little loss. Nothing makes firmer or better pork and lard than peas, and the manure from pea-fed pigs is exceedingly rich. A heavy crop of peas, too, is a capital crop to precede winter wheat. They will smother the weeds, and, if sown early, are off the land in good season to allow thorough working of the land before wheat sowing. If other food is scarce, a few of the peas may be cut in June, as soon as the pods are formed, and fed green to the pigs, and a daily allowance may be fed until the peas are fully ripe. In fact, many farmers feed all their peas to the pigs without thrashing. But this is a wasteful plan. When the peas are ripe, pigs will do much better on them cooked, or at least soaked in water for twenty-four hours before feeding. And in addition to this advantage, pea straw, when well cured and carefully harvested, is nearly as good for sheep as clover hay, and certainly will much more than pay the expense of thrashing. A large farmer in Michigan, who has made himself and his farm rich, attributes his success principally to growing a large quantity of peas every year, and feeding them to pigs. He thrashes the peas, and cooks them, but does not grind them, as he thinks cooking is better and cheaper than grinding. The manure from his pea-fed pigs has made his farm one of the most productive in the State.

FATTENING PIGS NEAR LARGE CITIES.

Nurserymen, seed growers, and market gardeners near our large cities require great quantities of manure. Hitherto they have obtained it from the horse and cow stables in the city, but the demand is greater than the supply, and the price is so high that many are looking to other sources for manure. In Rochester, the price of manure from the stables is $1.25 per load, and by the time it is well rotted, it requires three loads of fresh manure, as drawn, to make one load of rotted manure, as applied to the land. This, added to the expense of drawing, brings the cost of the manure up to about $100 per acre. In Geneva, N. Y., where the nursery business is carried on very extensively, the price of manure is even higher still, or $1.50 per load at the stables. And there, as well as at Rochester, some of the nurserymen are turning their attention to fattening sheep in winter for the purpose of obtaining cheaper and better manure. The result, so far, has been eminently satisfactory where the nurserymen have land enough to raise their own clover hay.

But where land is very high, and where, consequently, it will not pay to raise clover hay, some other system must be adopted. Pig feeding would seem to offer the best prospects of producing the richest manure at the least cost.

For this purpose, the first requisite is a good breed of pigs, that will mature early, and fatten at any age, so that they could be disposed of at any time when choice fresh pork was in demand, at good prices. Unfortunately, such pigs are difficult to find, and will continue very scarce until farmers learn the importance of using none but thorough-bred boars, of a highly refined breed, with properly selected common sows. With young pigs, so bred, we have no doubt that the system of feeding pigs on purchased food might be profitably adopted near our large cities. Certainly, manure could be obtained in this way at far less cost, in proportion to its value, than is now generally paid for it. A study of the table on page 139, showing the value of manure from different foods, and an examination of the results of Lawes' and Gilbert's experiments in feeding pigs with different foods, showing what kinds produce the greatest increase, will enable any one to select feeding stuffs with judgment and economy. Three things have to be considered: the cost of the food; its feeding value, and the value of the manure obtained from its consumption. We have given all the data necessary to enable any intelligent man to engage in this business with confidence and success. If there is any error, it is on the safe side, for we are satisfied, from our own experience, that well-bred pigs can be so fed as to give a greater increase from the food consumed than was obtained in Mr. Lawes' experiments, when no special attention was paid to the breed. In this connection some useful hints may be obtained from the following chapter.

Fig. 52.—BROOD SOW, BELONGING TO THE DUKE OF BUCCLEUCH.—*Drawn by Landseer.*
From Stephens' Book of the Farm.

CHAPTER XIX.

ENGLISH EXPERIENCE IN PIG FEEDING.

In some respects, the farmers of England and the farmers of the Middle and Eastern States are similarly situated. England does not raise scarcely half as much wheat as is needed by her population, and the same is true of our Middle States; while in New England, enough wheat is not raised to support one-tenth of the population. English farmers are thrown into direct competition with the produce of all other countries, and the farmers of New England and the Middle States have to compete with the produce of the Western States. Prices depend less on the home crop than on the yield in those countries from which the principal supply is derived. A poor crop at home is not necessarily compensated by higher prices. And, therefore, it is particularly important to guard as much as possible against poor crops from unpropitious seasons. High farming is found to be the best safeguard. But high farming not only requires thoroughly drained and well tilled land, but abundance of manure. English farmers must compete with the cheap land of our Western States, and also with the cheap labor of Ireland and the continent. But, in spite of all this, they continue more prosperous, as a whole, than the farmers of any other country.

We cannot adopt the English system of agriculture, but the principles on which it rests are as applicable here as there. What the farmers of New England and the Middle States require, is more capital, more labor, and more manure. And, in many places, manure can be obtained cheaper and better from feeding well-bred pigs than in any other way. This, at any rate, has been the experience of many English farmers, and the prospects are still more favorable in the New England and other Atlantic States, because food is cheaper than it is in England, and the large cities are not as well supplied with choice fresh pork as are those of England, and consequently it brings, or would bring, if it could be obtained, a relatively higher price, as compared with beef, mutton, and barreled pork.

In 1862, Mr. Baldwin, of Breton House, near Birmingham, delivered a lecture before the Worcestershire Agricultural Society on the breeding and feeding of pigs, in which he said:

"In 1845, he entered upon a farm at Kingsnorton. In 1846, he purchased two gilts and a boar, of the Tamworth breed, and although he began breeding with only three pigs in 1846, in 1851 he sold £1,000, say $5,000, worth of store and fat pigs within one year; and in the years 1852, 1853, 1854, and 1855, he sold about £1,000 worth each year. The idea of feeding such numbers of pigs was first conceived by him at a county meeting at Worcester, in 1849, after free trade had come into full operation. One of the speakers produced many samples of foreign produce at amazingly low prices. Among them was a good sample of Egyptian beans, at 9s. and 9s. 6d. per bag; Indian corn at the same price, and Dantzic wheat, also, very low. 'Gentlemen,' exclaimed the speaker, 'can you grow them at these prices?' He (Mr. Baldwin) looked on the bright side of the question, and began to ask himself how he might turn the low price of grain to good account. It struck him that, as he had a great many store pigs, he would feed them instead of selling them as stores. He accordingly bought a large quantity of Indian corn, at from 9s. to 9s. 6d. per bag, [200 lbs.], to begin with; and within two years and a quarter from that time, he bred, fed, and sold £2,000 worth of pigs, and cleared, after paying all expenses, £500, besides making a vast amount of manure, which he considered far better than guano, because more durable.

"The plan which he adopted in breeding was, to put the sows to the boar in November, and pick the breeders principally from the earliest pigs, when he got his stock up to about forty breeding sows. In picking the breeders, he used to pick them several times over, as it frequently happened that those which looked the best and prettiest when young, altered considerably when they got three, four, and five months old. The rule was to pick long-growing pigs, and those that were straight and thick through the shoulder and heart, and experience had convinced him that his method of choosing was a correct one. He always kept to the Tamworth breeds, generally purchasing the boars, but breeding the sows. If he found the pigs getting too fine, he purchased a good strong boar, and if the animals exhibited tendencies the other way, he picked a boar of good, small bone, but was always particular to select a boar that was thick through the shoulder and heart, and a straight-growing pig, of the same color and breed. By carefully following this plan, he got the breed so good, that it was a rare occurrence to see even a middling pig in all the herd, though he bred from 250 to 300 each year. His plan of keeping was as follows: As soon as the sows littered, they were kept on kibbled [crushed] oats, scalded, with raw Swedes or cabbage; and when the pigs got to the age of three weeks or a month, he turned the sows out from them for a short time every day, and gave the pigs a few peas or Indian corn while the sow was away. When the weather was fine and warm, the pigs went out with the mother into a grassy field for a short time. He found that young pigs, from the age of three weeks, required dirt or grit; and, therefore, if the weather was bad, and they could not be turned out, it was necessary to put some grit into the sty. This was quite important, as he believed it was necessary for the proper digestion of their food. He had had young pigs looking very bad and drooping, but when turned out, that they might get dirt, they soon became all right again. In fact, it was absolutely necessary, during the whole life of a pig, to allow it an opportunity of getting grit or dirt, or it would not thrive well.

"At seven or eight weeks old, all the pigs he did not require for breeding he had cut, and began to wean them a fortnight afterwards. He then turned them into a grass field, with a hovel for them to run into, and allowed each pig a quart per day of peas, Egyptian beans, or Indian corn. He gave them one pint of the corn in the morning, and the other in the evening, with regularity as to time and quantity, and found it better to give it them on the grass, in a clean place, each time, than in a trough, as it prevented quarreling, and each pig got his share. With this quart of corn per day, and what grass they got during the seven months of the year, with nothing but water to drink, the pigs would, on the average, make 5 lbs. of pork, each, per week. After eight months, he allowed an extra half pint of corn per day. At the present price of corn (1862), the allowance would cost about 1s. per week [24 cents], for each pig; grass, 4 cents; attention of man, 2 cents; total cost, 1s. 3d. (30 cents), leaving a profit of 24 cents per week on each pig, when pork was 12 cents per pound; it was now 14 cents.

"One man attended—well, to from 200 to 300 pigs; he was an Irishman, for few Englishmen liked the job sufficiently well to take an interest in them, and carelessness on the part of the man materially decreased the profits.

"He kept the store sows, when with pig, the same as the other stores. They ran about in a field until a fortnight before pigging, when he placed them in a covered shed, so constructed as to admit as much sun as possible. Young pigs, kept in the manner described, were always nearly fat enough for porkers, and did not require more than two or three weeks feeding on meal. It was time enough to begin to feed pigs for bacon at eight or ten months old. Good breeding sows he allowed to have two farrows, and sometimes three, but never more, and then fed them for bacon, supplying their places with young sows.

"In selling store pigs, he charged a certain price per pound, and allowed the purchaser to pick the pigs from the field, which plan always gave satisfaction, and secured a return of custom. It was desirable, in breeding animals, to have as little bone as possible, in proportion to flesh. He had tested a cut sow of his breed, which weighed 640 lbs., and the whole of the bones, after the flesh had been boiled from them, weighed only 21 lbs., so that for every pound of bones there was 32 lbs. of meat. His pigs made 1 lb. of flesh for every 4 lbs. of good Indian corn, barley or pea-meal; as a rule, he preferred the Indian corn. He considered it always to be more profitable to feed good food than upon that of inferior quality. As a rule, pigs would thrive better for being turned out once a day, except in wet weather, and would also be healthier, more active, and have a cleaner appearance. One of the greatest pleasures his breeding afforded him was to see the number of laboring men who came to buy from him, and he hoped to live to see the day when every laboring man would have a good pig in his sty."

Mr. Baldwin's experience is the more valuable, as he seems to keep pigs to sell to the butcher, or to those who intended to fatten them. His success is not due to selling thorough-bred pigs at high prices for breeding purposes.

A Yorkshire farmer, who occupies 280 acres of land, half under plow, and half in grass, and who raises and feeds a large number of the small Yorkshire and Cumberland breed of pigs, writes Mr. Sidney as follows:

"I am a farmer, and I keep pigs for profit, and I have no stock that pays like them; but I have found a surprising difference in the feeding qualities of the different breeds, and I am not astonished at farmers saying pigs will not pay. I think the medium size pay better than the large bacon hogs. For eleven years I have kept an account of all my pigs cost, and what I sell, and at the year's end I know the truth. * * I spend $3,500 a year for purchased food, but little on any manure, except lime and salt. I make all the manure I can, and make it good. I calculate I get my pig manure free, but not my cattle manure. For the first fortnight the little pigs live upon the sow's milk. Then they will begin to eat a little dry wheat. As soon as they begin to eat freely, have a place where they can creep to feed, where the sow cannot get at their meat; and feed them separately, twice a day, with milk, meal, and bran, and once a day with dry wheat. But beware of over-feeding them, or any young animals. At six weeks old, the boar pigs are usually castrated, and at eight weeks old, the litter may be weaned by taking away the sow by degrees. But if the sow is not wanted to breed again directly, and you want to forward your pigs, it is a good plan to let them be with the sow, at night only, until they are twelve weeks old, and then they ought to be in very good condition.

"After twelve weeks, the treatment will depend upon what they are wanted for. If to be made the best of, feed them for the next twelve weeks on boiled meal, vegetables, and a little bran—two feeds a day—keeping about six together in a sty, warm, and well bedded. Keep them on *cooked food*, and a little meal every day, until within six weeks of being killed, when they should have as much barley-meal and water as they can eat. It is a waste of money to give them raw meal all the time, but they should always be gaining until the slaughtering day —to go back is a loss."

It would seem that the plan this farmer adopts, or at least that which he considers best, varying in practice probably with the demand for fresh pork, is to push his pigs forward as rapidly as possible, and sell them when six months old. And this is the system which, in the neighborhood of our large cities, we believe, will be found the most profitable in the United States. For this purpose we unquestionably require pigs of some of the small breeds, that will mature early.

A dairy farmer, who keeps Berkshire pigs, says: "My stores, farrowed in March, are fatted off by December, making from ten to twelve score, although I have often had them much heavier. Pigs of this weight are always more salable in the London Newgate Market, at sixpence or a shilling a score more than heavier ones. I have grown a pig of the Berkshire breed over 40 score (800 lbs).

"Second litters, coming in about December, at three months old, will do for pork. The sow will then be in again in March or April.

"The whey runs from my dairy into a vault near the piggery, in which I have large bins to mix the whey and meal together, allowing it to ferment for three days before using it. If I am well off for roots, I have a good quantity pressed, steamed, and minced with whey and barley-meal. In the winter, a few beans or lentils, ground. *If convenient*, give warm food. Have not more than six pigs together. Warm sties, clean, and the pigs well groomed with brush and linseed oil, which will cleanse the skin, and kill the lice with which they are often annoyed."

Another pig feeder recommends pulping roots, leaving them to ferment for thirty-six hours, and then mixing the pulp, by alternate spadefuls, with meal. This he thinks as good as cooking, and much cheaper.

He does not mention the kind of roots used, but mangel wurzel, beets, and parsnips, are best adapted to our climate and circumstances. With rich land, and good culture, a large amount of nutritious food can be obtained per acre, and feeding them out to pigs, with meal, will make very rich manure, and thus we obtain the means to raise more food, and keep on increasing the productiveness of the land.

A Yorkshire pig breeder says: "I have had a great, many York-Cumberland pigs that gained—

7 lbs. each, per week, up to ten weeks old.
10 lbs. per week for the next seven weeks.
14 lbs. per week until they weighed 23 stone.

"I can put on 18 lbs. a week until a certain time, and then they begin to put on less and less every day, until at last you feed at a loss. The pig should be killed when the point of profit for daily food is turned. For this reason the pig should be weighed weekly.

"After trying nearly all the different kinds of cereals, and weighing my pigs once every fourteen days, I have come to the conclusion, *if you want to gain weight fast*, give plenty of barley-meal and milk; *if you want to make the most of the food consumed*, give boiled vegetables and boiled meal, and finish off with raw meal.

"On the first plan, time is saved at the expense of food consumed. On the second plan, time is lost, and the food saved."

If by "food" is meant *meal*, the statement is probably correct; but that we ever save food, absolutely, by feeding slowly, is a proposition that has never been proved, and is contrary to sound theory and the general experience of the best feeders. A fattening animal should certainly have all the food it can digest and assimilate. To keep him on short allowance is to waste both time and food.

Another correspondent of Mr. Sidney writes: "With tolerably good land, and no lack of capital, a farmer cannot do better than cultivate white crops alternately, and, with a moderate dairy, confine his stock exclusively to pigs. Let him consume his oats, sell off both wheat and barley, and buy Indian corn and bran. Indian corn is about the same price as barley, but sixty, instead of fifty-two pounds to the bushel. A bushel of barley-meal is generally supposed to add 10 lbs. to the weight of a pig. I have found, in my latest experiments, that a bushel of Indian corn produced an increased weight to a pig of 15 lbs.

"Indian corn," says Dr. Voelcker, "is richer in fat-forming matters than almost any other description of food. The ready-made fat in corn amounts to from five and a half to six per cent. But animals should not be fed exclusively on Indian corn, because the flesh-forming matter in it is small. Bean-meal [or pea-meal] supplies the deficiency. Five pounds of Indian corn, ground or crushed, to one pound of bean-meal [or pea-meal], is a mixture which contains the proportions of flesh-forming and fattening matters nicely balanced."

Another Yorkshire farmer writes: "We are now (1860) fattening pigs on wheat costing $1.20 per bushel [in gold], which, as large bacon pigs are selling at 12 cents per pound, leaves a handsome profit for fattening, even at the present high price of stores.

"But," he adds, "the farmer who is wise, will keep both these profits in his own hands. He will rear his own stores, and grind up his own grain for feeding them. If he wants pigs to pay, he does not starve them for twelve to eighteen months, leaving them to roam about the fields, consuming as much food among twenty as would feed thirty, rooting and turning over a fold-yard dung heap; but he finds, with the corn, that it will cost him in money half its feeding value, and gets the manure into the bargain.

"A well managed pig-feeding establishment, near any great town, ought to pay in times of low-priced grain. Unlike beef and mutton, every inch of a pig is in demand, and the offals are sold at good prices as dainty bits."

We might quote much other evidence of a like character, but the above is sufficient to show that the English farmers can send to the United States for Indian corn, pay freight, commission, and expenses, and then use it at a profit in fattening pigs, which are sold at prices no higher than the same quality of pork brings in New York, Boston, or Philadelphia. Cannot we do the same thing here? Let those who undertake it, however, remember that the demand is for choice, fine-boned, well-fatted pigs, of the best quality. Such pigs would bring from three to five cents per pound more than common hogs, and this, in itself, is a large profit.

CHAPTER XX.

LIVE AND DEAD WEIGHT OF PIGS.

The three grade Essex pigs (Nos. 3, 4, and 5) in Dr. Miles' experiments at the Michigan Agricultural College, previously alluded to (see page 118), were killed when 31 weeks old. Their live and dressed weights were as follows:

	Live.	Dressed.	Dressed to Live Weight. Per cent.
No. 3	135½	112½	83
4	156	132½	85 nearly.
5	145½	122	83¾

The live weight was taken *before feeding.* For such small pigs, this shows a very high proportion of dressed to live weight.

An Essex pig, about fifteen months old, belonging to the writer, weighed, after sticking, 445 lbs., and dressed, as weighed the next day by the butcher, 409 lbs.—a shrinkage of only a little over 8 per cent. Allowing 10 lbs. for the blood, the pig would have weighed, alive, 455 lbs., and dressed nearly 90 per cent.

We have no doubt that a highly refined pig of any of the small breeds, well fed during its whole growth, and thoroughly fattened, will shrink less than 10 per cent on its fasted live weight.

Messrs. Lawes and Gilbert accurately ascertained the live and dead weight of the fifty-nine pigs on which their experiments, previously alluded to, were made. The actual, average live weight, after fasting, of the whole fifty-nine pigs, was 212³/₄ lbs., and the average dressed weight, 176 lbs., 5.3 oz., or a little over 82¹/₂ per cent.

The following table shows the actual average weight of the *different parts* of these fifty-nine pigs, and in the right-hand column we give the per centage weights:

TABLE SHOWING THE WEIGHT OF DIFFERENT PARTS OF A PIG WEIGHING, ALIVE, 212¾ lbs. (AVERAGE OF 59 PIGS.)

	Actual weight.	Per cent.
Stomach and contents	2 lbs., 10.4 oz.	1.28
Caul fat	1 " 2.3 "	.54
Small intestines and contents	4 " 8.4 "	2.20
Large " " "	8 " 5.7 "	4.04
Intestinal fat	2 " 5.6 "	1.06
Heart and Aorta	0 " 9.6 "	0.29
Lungs and Windpipe	1 " 9.1 "	0.76
Blood	7 " 10.1 "	3.63
Liver	3 " 4.5 "	1.57
Gall-bladder and contents	0 " 2.1 "	0.06
Pancreas ("sweetbread")	0 " 6.6 "	0.19
Milt, or Spleen	0 " 4.7 "	0.14
Bladder	0 " 2.5 "	0.08
Penis	0 " 7.1 "	0.21
Tongue	1 " 0.2 "	0.48
Toes	0 " 2.9 "	0.08
Miscellaneous trimmings	0 " 8.8 "	0.26
Total offal parts	35 " 4.6 "	16.87
Carcass	176 " 5.3 "	82.57
Loss by evaporation, etc.	1 " 2.1 "	0.56
Live weight after fasting	212 12 "	100.00

For the sake of comparison, we may say that the average, of 249 sheep, killed at Rothamstead, by Messrs. Lawes and Gilbert, was, fasted live weight, 153 lbs., 10.2 oz.; Carcass, 91 lbs., 12¹/₂ oz.; Per centage of carcass to live weight, 59³/₄. The sheep were Cotswolds, Leicesters, Hampshire and Sussex Downs.

The mean fasted live weight of 16 heifers and steers, killed and slaughtered at Rothamstead, was 1,141 lbs.; Carcass, 680³/₄ lbs.; Per centage of carcass to live weight, 59.31. In other words—

A moderately fat heifer or steer will dress................................59¼ per cent.
A moderately fat mutton sheep will dress.......................... ......59¾ " "
A moderately fat pig will dress............................ ....82½ " "

The lightest of Mr. Lawes' half-fattened and fattened pigs dressed a little less than 74 per cent, and the heaviest over 87³/₄ per cent.

CHAPTER XXI.

BREEDING AND REARING PIGS.

The point of first importance in breeding pigs is the selection of the boar. In raising thorough-bred pigs, of course we must have a boar of the same breed as the sow. This remark may seem superfluous, but we have met with ordinarily intelligent men who thought that a boar, descended from a thorough-bred Cheshire sow, got by a thorough-bred Chester White boar, was thorough-bred. And we have known a farmer who put a Chester White sow to an Essex boar, speak of all the white pigs in the litter as Chester Whites, and all the black ones as Essex. Thorough-breds must be descended from thorough-breds, and both parents must be of the same breed.

But in raising pigs for the butcher, we are not confined to any particular breed. Our selection of the boar must be made in reference to whether the pigs are to be fatted and sold at a few months old for fresh pork, or whether they are to be kept until they have nearly attained their growth before being fattened. Reference must also be had as to whether we wish large hogs, or smaller and finer ones at a less age. Much, too, will depend upon the sow we wish to breed from.

Defective as the majority of our pigs are, there are, nevertheless, few sections where we cannot find some strong, vigorous sows, of good size, suitable for crossing with the improved breeds. This is especially true where the Chester County pigs have been introduced. We could not ask for better sows to start with than a grade Chester County sow. It is an easy matter to find strong, vigorous sows, of good size, in any neighborhood where the Chester County or similar large breeds have been introduced.

If a farmer wishes to keep hogs until they are from fourteen to eighteen months old, letting them run in the barn-yard the first winter, and in a clover pasture and stubbles the next summer, and to be fattened in the fall, he cannot go wrong in selecting a large, vigorous, somewhat coarse sow, showing more or less Chester County blood. Then put her to either an Essex, Berkshire, Suffolk, or Small or Medium Yorkshire boar. We think it matters comparatively little which of these breeds is used, provided, always, that they are good specimens of the breed, and *are thorough-bred*. Better pay five dollars for the use of a thorough-bred than accept the service of a grade or common boar for nothing.

If the sow has had pigs, say the middle of March, they may be weaned in six weeks; and if the sow has been properly fed, she will take the boar in a few days after the pigs are weaned. We should then get one litter of—say grade Essex—about the first of September. The sow, during the summer, should, if possible, have the run of a clover pasture; and, if she is not in good, thriving condition, with this, and the wash or milk from the house, throw her two or three ears of corn a day. She should not be too fat, but there is not one farmer in a thousand who ever falls into this error. Let her have plenty of exercise, and if she is fully *half fat* by the time she comes in, all the better. If she is a good mother, nearly all her accumulated fat will find its way to the little ones in the milk before they are six weeks old.

For two or three weeks before she is expected to farrow, let the sow be put in a pen by herself at night, so that she may become accustomed to it. She may be allowed to run out during the day, but should always be fed separately in the pen, and in this way she will soon come to regard the pen as her own, and will go in as soon as the door is open. Let no harsh word be spoken, or a kick or a blow, on any provocation, be resorted to.

The pen should have a rail around the side, about six inches from the floor, and eight or ten inches from the sides of the pen, so that if she makes her bed near the sides of the pen, as she almost invariably will, the rail will afford a space for the little ones to slip under, and thus prevent their being crushed against the sides of the pen. As, at this season, the weather is warm, she will need but little straw. The better plan is to put two or three times as much straw as is needed into the pen a week or ten days before she is expected to pig. By lying on it she will make it soft, and this is very desirable. If any of it becomes wet or dirty, remove it from time to time when the sow is out. As the time approaches, she will select a particular spot, and "make a bed." When she is eating, or out of the pen, examine the bed, and see that the sides are not too hard, or compacted together too closely, and that they are not more than four or five inches high; if so, remove a little of the straw. It is better to have too little than too much. After this, the sow should be left to herself. With gentle thoroughbreds, that are accustomed to being petted, we keep a close watch during such an interesting event, rendering assistance if necessary; but, as a rule, and especially with common pigs, it is far better to trust to nature, and let things take their course. At this season of the year, especially if the sow has had the run of a pasture, and is in a thrifty condition, there will seldom be any trouble. The little pigs will come strong, and commence to suck in a minute or two after they are born. On no account disturb the sow until all is over. This may be two hours, and sometimes longer. Do not be in any hurry to feed her. But when she gets up, let her have all the milk or slop that she will drink. It is better to watch her, and keep pouring it into the trough as long as she will drink it up clean. Let her have all she can drink, but leave none in the trough. We are aware that these directions are not in accordance with the general rules on this subject. There are those who think that the sow should be kept on short allowance, so that she may be wide-awake, and quick to hear the scream of any of the little ones she may be lying on. This is all very well, but the chief danger occurs from the sow getting up and lying down again; and if she has a good meal, and eats it all up clean, she will be more likely to lie still during the night than if she is hungry. After she has eaten, and when she goes back to her bed, you will be there to hear if she lies on any of the pigs, and can go to the rescue. When she has once lain down, there is little danger until she gets up again. If all goes well for the first two nights, there will rarely be any loss or trouble afterwards.

Give the sow all the milk or slops she will drink, but little or no grain for the first week or ten days. If the little pigs scour, change the food of the sow. There is nothing better for her than skimmed milk, not too sour, and the next best thing is two quarts of fine middlings, scalded with two or three quarts of boiling water, and the pail afterwards filled up with water sufficient to cool it to the temperature of new milk. And here we may say that some men do not seem to know how to scald bran or meal properly. We have seen them put the meal in the pail and pour on the water, and then fill up with cold water at once, and without previous stirring. The proper way is to put—say two quarts of the bran or meal into the pail, pour on the boiling water, and stir it up until every particle is wet or moistened; and the longer it remains before the cold water is added, the better. The object is to soften and cook it, and make it more easily digestible. When properly prepared, it should look like fresh milk. Do not say that the pigs will not pay for all this trouble. It is a great mistake. In the first place, it is very little trouble, and in the second, the future growth of the pigs depends very much on their being well cared for while young.

When the pigs are two weeks old, a little shallow trough should be made for them. Nothing is better for this purpose than three or four feet of a tin eaves trough, turned up at the ends. Nail it to the floor, so that the pigs will not upset it; and, if possible, put it where the sow cannot get at it. Then put in half a pint or so of sweet milk. Let them drink and waste what they will of it, but always clean it out before fresh food is added. Try to teach them early to eat their meals promptly, and then lie down to sleep. Give them a small handful of oats, or, better still, three or four tablespoonfuls of oat-meal, increasing the quantity daily, but never giving more than they will eat up clean. If fed too much at one time, and too little at others, it will produce scours, and retard the growth of the pigs. At three weeks old, a litter of eight or ten pigs will eat a quart of good oats four times a day. They seem particularly fond of cracking the oats and eating out the kernels.

After the first week or ten days the sow should have richer food—say two quarts of fine middlings, and a quart of oat or corn-meal, three times a day. Let her have all she will eat, and in a week or ten days later, give richer food. Boiled barley is excellent, but it is better to vary the food, so as to induce the sow to eat more. We often throw our sows an ear or two of corn after they have eaten their regular meal. The more food the sow can be induced to eat, the richer will be the milk, and the more rapidly will the little pigs grow.

When about six weeks old, the pigs should be altered. Do not be tempted to reserve one of them for a boar. No matter how handsome and well formed he may be, it is absolute folly to use him for breeding purposes. Select out one or two of the best sows, but alter all the boars. The sow pigs will grow and fatten more rapidly if spayed, but it is not often, in this country, that we can find men who are able to perform the operation with safety. Where there are such, all the sow pigs not intended for breeders should be spayed a week or ten days before weaning. There is nothing better to apply to the wound than petroleum—not kerosene—but the crude oil.

The time of weaning will depend on the time when it is desired to have the next litter of pigs. If the sow is in good condition, she will take the boar in a week or two after the pigs are weaned. And if the sow and pigs are well fed, the pigs may be allowed to remain with the sow until ten weeks or three months old, if there is time enough for the next litter, and the sow is strong enough to stand the drain on her constitution. If she is not strong, wean the pigs when six weeks or two months old.

It is better not to remove all the pigs at once; or, if this is done, let them return to the sow for a few minutes at the expiration of twelve hours, and again at the expiration of twenty-four hours. We prefer, however, to let one or two of the weaker pigs remain with the sow for a week or so after the others are removed.

At the time of weaning, the pigs should have extra attention. Feed them five times a day—the first thing in the morning, and the last at night. If they have all they can eat, they will not pine for the mother. Nothing is so good for them as milk. A little flax-seed tea, oat-meal gruel, or corn-meal gruel, mixed with the milk, or given separately, will be good, and acceptable. As the weather, by this time, is getting cold, it will be well to give warm food. But guard against giving it too hot. It should not be warmer than new milk.

There is, perhaps, nothing better for the pigs than corn pudding and milk. Put two quarts of corn-meal into a pail, and pour on two or three quarts of boiling water, and stir it until all the meal is wet, then fill up the pail with milk. But be very careful that the scalded meal is all mixed with the milk. It often happens that there will be lumps of meal hot enough to scald, although the milk surrounding it is only warm. Such lumps should be broken up and mixed with the milk before feeding to the little pigs.

We need hardly add that all pigs should be allowed a constant supply of fresh water. There are few things of more importance in the management of pigs.

Let the pen be warm, clean, and well ventilated, but with no cracks for the wind to blow through on to the pigs. And, above all, let the pen and bedding be *dry*. There should always be litter enough for the pigs to bury themselves in. Warmth, to a certain extent, is equivalent to food, and, what is of more importance than the saving of the food, it *saves digestion*. Let the pigs have all the exercise they wish, and then do not be afraid that warm, dry, clean, and comfortable quarters, with abundance of wholesome food, will make them tender. We are aware that this is a common idea, but it is an erroneous one. A cold wind or storm, that will send a half-starved and neglected pig squealing around the barn-yard, with hair on end, head down, and back up, will have no effect on pigs treated as we have recommended. And there is nothing more important than to have young pigs in a healthy, vigorous, almost *fat* condition, before winter sets in.

The pigs are now three months old, and should weigh 75 lbs. to 80 lbs. each. We have had grade Essex and Berkshires (which are not as large as grade Essex and Chester Whites) that weighed 88 lbs. when three months and four days old.—And it should be remembered that, during two months of the time, the pigs get most of their food from the sow; and during the next month, they eat far less food than older pigs.

During the winter, the pigs may be allowed the run of the barn-yard, to pick up what they can find. If the cattle are fed grain or oil-cake, a certain number of pigs will keep in good condition on the droppings of the cattle, and on food which would otherwise be wasted. Let the young pigs, however, have a separate pen from the old ones, and see to it that they have enough food to keep them in good condition. By throwing them an ear or two of corn in the pen, they will soon learn to be ready at the appointed time to enter the pen for the night, without trouble. On no account let them go to bed hungry. Let their stomachs be well filled—say at five o'clock in the evening—and they will sleep quietly until eight o'clock the next morning. In fact, a well-bred and well-fed pig will sleep three-fourths of his time, during the winter. If not disturbed, and tempted with fattening food, he will eat little and gain little. And sometimes, like other hibernating animals, he will live on his own fat.

As spring approaches, the young pigs will need more food, and fortunate is that farmer who has a liberal supply of parsnips, sugar-beets, or mangel wurzel for them. These roots, pulped or rasped in a cider-mill, mixed with a little corn-meal, are a cheap and excellent food for pigs in the spring. But, whatever the feed, let the pigs have all they need to keep them in a good, thriving condition.

As soon as the clover is fairly growing, the pigs should have the run of the clover pasture. They will get three-fourths of their food in the pasture, and we need hardly say that, where clover grows as abundantly as it does with us, it is the cheapest food that can be fed to a pig. With clover, and the slops from the house and dairy, the pigs will keep in a thriving condition, but it is a waste of time and food to depend on this alone, with pigs intended for the butcher. If fed from a pint to a quart of corn, or corn-meal, a day, they will eat just as much clover, and will grow nearly as fast again. After harvest, they will pick up considerable food on the grain stubbles; but if as fat as they should be by this time, stubble gleaning can be more profitably left to the breeding stock and spring pigs.

By the first of November, such pigs as we have described, fed as here recommended, should be in prime order for the butcher, and can be sold at any time when the price is satisfactory.

They should average 400 lbs., dressed weight. The pork is of the highest quality, and the lard keeps firm and hard during the hottest weather in summer, and makes excellent pastry.

REARING AND MANAGEMENT OF SPRING PIGS.

Spring pigs, intended to be fattened and sold when about nine months old, should come early in the spring, and should have the best of care and feed. A warm, dry pen, is absolutely essential. Thousands of pigs are lost every spring for want of a little forethought in making the pen ready for the sow to litter in. In a properly constructed pen there is little to be done, except to clean it out a week or ten days before the time the sow is expected to pig, and provide a liberal allowance of dry straw. It is not well to have too much straw in the pen at the time of pigging; but, as already explained, straw which has been in the pen for a week or so is softer and better than fresh straw. We would place straw in the sleeping apartment to the depth of a foot, and then remove the wet or soiled portions daily until, by the time the sow pigged, there would not be more than is needed to keep the mother and little ones warm. Two or three inches of soft straw on the bottom of the pen, under the sow, will be trod firm, and act as a non-conductor of heat, and will not increase the danger of the sow lying on the pigs. The danger arises from having too much *loose* straw in the pen, and from having the sides of the bed too high and firm.

It often happens that the pen in which the sow is placed is ill-adapted for the purpose. In this case, some temporary expedient for keeping out the cold winds must be resorted to. If nothing better can be done, every hole and crevice can at least be stopped with straw. The principal danger is during the first few hours after the pigs are born. If they can be kept warm and safe for two or three days, there is little danger of losing them. But for health and thrift, it is very desirable that they never be exposed to cold storms; and what is of even still greater importance, the pen must always be *dry*.

We would again endeavor to impress on our readers the importance of attending to these matters in advance. Few things are more vexatious than to lose a nice litter of pigs for want of half an hour's time in making the pen dry, warm, and comfortable. If we lose a calf, we have still the milk of the cow, but if we lose a litter of pigs, there is no compensation. It is a dead loss of what the pigs would have been worth when a month old.

We have said that for fall pigs, to be kept fourteen or fifteen months before killing, there are no better pigs than those obtained from a Chester White sow, put to a thorough-bred Essex, Berkshire, or Small Yorkshire (Suffolk) boar. But for spring pigs we need a little more refinement. They should be three-quarters Essex, Berkshire, or some other fine breed—that is to say, a sow from the first cross of Essex and Chester White should be put to an Essex or Berkshire boar. This would give a highly refined, small-boned pig, that would mature early, and fatten rapidly. During the summer, however, they will require better food than the older and stronger pigs. They should have the run of a clover pasture, but should be favored in the distribution of the milk, and should have, in addition, sufficient grain of some kind to keep them fat enough for fresh pork at all times.

It often happens that the most profitable way of disposing of such spring pigs as here described, is to sell them when three, four, five, or six months old for fresh pork. We have sometimes thought that butchers do not make sufficient difference in the price of such pigs as compared with common pigs. In fact, butchers have said to us: "All that you say is true. These pigs make splendid pork, but our customers will pay no more for it than for common pork, with half as much again bone in it." The truth of the matter seems to be this: There is not enough of such pork sent to market to establish the grade. Few people know that there is as much difference between the pork from a four-months-old, well-bred, and well-fed pig, as compared with an eight-months-old, ill-bred, and ill-fed pig, as there is between a sirloin and a round steak. In Boston, a sirloin steak is now (March, 1870) quoted at 36 cents and 38 cents per pound, and a round steak at 20 cents and 25 cents; chuck rib at 12 cents and 15 cents, and soup pieces at 5 cents and 8 cents per pound. Here is certainly difference enough to stimulate us to improve the form of our animals. Let farmers furnish *good* fresh pork, and there will be found those who are willing to pay a liberal price for it. At any rate, if the pigs are kept in high condition, they will be ready at all times for the butcher; and if the price is suitable, they can be disposed of, and if not, they can be kept until nine or ten months old, and sold for fat pork. Spring pigs should never be kept on short allowance. It is almost impossible to keep them too fat. To keep them in a half-starved condition until the corn crop is ripe, and then shut them up to fatten, is a very expensive way of making pork. We have known a lot of spring pigs kept in this way, by a farmer who seemed to fear that, if he fed a little corn during the summer, his pigs would not "grow," that were shut up to fatten in October, and fed soft corn at first, and afterwards sound corn in the ear, all they would eat, that did not, when killed in December, average 100 lbs. each, dressed weight. A well-bred pig of the same age, well-fed from the day he was born, (and before,) would have dressed 300 lbs.

CHAPTER XXII.

MANAGEMENT OF THOROUGH-BRED PIGS.

The first object in the management of thorough-bred pigs is to secure perfect health. If any animal manifests the slightest tendency to disease of any kind, it must be rigorously rejected. Moreover, if in a litter of pigs there are any defective animals, we would fatten the sow and dispose of her. It is not safe to breed from her. And if the same defect manifests itself in the litters of other sows, bred to the same boar, it is pretty conclusive evidence that the boar is not perfectly sound, and he should be at once rejected. No matter how apparently healthy the parents may be, if there is any tendency to disease, or defects in form in the offspring, the probabilities are, that there is some latent disease in the parents; and even though we breed from none of their offspring but those apparently sound, yet we are never sure that the disease will not manifest itself in the next generation.

Next to health, the digestive and assimilating power of a thorough-bred pig is of the greatest importance. Without good digestion, rapid growth is impossible. The pig must have a stomach capable of extracting the nutriment from a large amount of food, and the process of assimilation must proceed with equal rapidity. These qualities are, in a good degree, under our control. In a thoroughly established breed, "like begets like," not only in form and color, but also in those qualities which determine rapid growth, early maturity, and a disposition to fatten easily. Check the growth of a young boar and sow by keeping them in cold, wet pens, on short allowance, and, though they themselves may afterwards apparently recover from such treatment, the evil effects will be seen in their offspring. They may be perfect in form, but they will not possess the maximum capacity of growth and fattening qualities. In the management of thorough-bred pigs, this idea must never be absent from the breeder's mind. So far as is consistent with health, the young pigs must be daily kept in such a way as to secure a rapid growth. All thoughts of "hardening" them by exposure to cold storms must be abandoned. All attempts at starving them, in hopes of making them more healthy and vigorous, must be given up—first, because it will not accomplish the object, and secondly, because if it would, we should lose one of the first objects we have in getting an improved breed of pigs—the capacity of converting a large amount of food into flesh and fat.

It has been supposed that the success of a breeder depends almost entirely on his judgment in selecting a male adapted to correct any deficiencies in the form or qualities of the females. But while this is sometimes very important, yet the real skill of the breeder of thorough-bred pigs consists, in great part, in his ability to keep his young pigs growing to their utmost capacity, and, at the same time, keep them in perfect health, and in good condition for breeding at the proper age. Let no farmer expect to succeed as a breeder of thorough-bred pigs if he leaves them to the care of an ordinary hired man. He must give them his own personal attention. If he objects to this, if he has no liking for a refined, well-bred, well-behaved, well-formed pig, let him turn his attention to some other business. It is, of course, not necessary that the owner should clean out the sties, or cook the food, or wash the pigs, and feed them. But he will find it of great advantage to know how to perform all these operations. Ordinary farm men have been so accustomed to let pigs wallow in the mire, and take care of themselves, that it is very difficult to get them to realize the importance of cleanliness, regularity in feeding, general kindness, and constant attention. It is not an easy matter to induce a common farm man to groom a horse thoroughly, and it is still more difficult to get such a man to clean a pig. And yet a breeder of thorough-bred pigs will find few things more important for health and for rapid growth, and for the development of the best points, than washing in summer, and cleaning them with a brush in winter.

An extensive range is almost as important for thorough-bred pigs as it is for poultry, and we think it a mistake for a breeder to keep more than one breed on the same farm. It is not only convenient and economical to let the pigs run out in a pasture during the spring, summer, and autumn, and in the barn-yard during the winter, but it is desirable for their health and vigor. It is not always easy to accomplish this object, even when one breed only is kept, and it must be still more difficult when two or three breeds are kept.

To keep up and improve the quality of the stock, it is absolutely essential to "weed out" all that show any tendency to deterioration; and on this account it is desirable to have a good-sized herd to select the breeding stock from. We must have at least two boars of each breed; and where two or three different breeds are kept, this is no slight expense. We would, therefore, earnestly recommend breeders to confine themselves to one breed.

THE BOAR.

A young boar must never be stinted in food. Until he is a year old, he should be kept growing as rapidly as possible, consistent with health and vigor. But at the same time, he must not be allowed to get too fat. We would let him have all the food he will eat. If he gets too fat, reduce the *quality*, but not the quantity, of the food. It is here that judgment and experience are particularly important. A person who has kept none but common pigs is very apt to think that his thorough-bred boar is getting too fat. The roundness and symmetry of the body, with the comparatively small growth of bone and offal parts, leads him to suppose that the pig is not growing fast enough. This is particularly the case with the small breeds. He thinks they are fattening inside, but are not growing; and, in order to make him grow, or, at all events, to prevent him from getting too fat, he turns him to a straw stack, or shuts him up in a pen, and feeds him nothing but dish-water and a few potato parings. Nothing can be more unwise. If the pig *is* getting too fat, which, in the case assumed, is not probable, the better plan is to turn him into a clover lot, or into a stubble field. What he needs is exercise and abundance of plain food. If it is winter, let him have less concentrated food, but give him all of it that he will eat up clean, twice a day. A few boiled potatoes and coarse bran, or bran alone, fed moist, makes a good winter diet in such a case. But so far as our observation extends, where one pig is injured from over-feeding, ten are stunted in growth from want of a regular and abundant supply of appropriate food.

At eight or nine months old, a boar of the small breeds, if kept in the way we have recommended, will have nearly completed his growth, and may be allowed to serve a few sows. But be careful not to let him have so many as to reduce himself materially in flesh, or check his growth. One service is sufficient for a sow, and to allow more is a mere waste of the strength and energies of the boar, and is probably injurious to the sow. To let a young thorough-bred boar serve a number of common sows, at a dollar a head, is mere folly. The English breeders usually charge a sovereign.

When the boar has attained his growth, he will not require as rich food. He should, however, have enough to keep him in perfect health and vigor. He should always have enough to fill his stomach. Bran and roots, or green clover, will ordinarily keep him in good condition. But when he is in active service, he must have richer food. In regard to the number of sows a full grown boar should be allowed to serve, it must be remembered that the proper season for having the sows come in is comparatively limited. From the middle of October until the first of December the boar is most in demand, and at this time, if full grown, may be allowed to serve twenty or twenty-five sows, and the same number during the spring season.

If the boar is a very valuable one, and it is intended to keep him for several years, he should be restricted to fewer sows—say eight or ten in a season. On the other hand, a boar that we intend to alter and fatten as soon as the season is over, may be allowed to serve all the common or grade sows that his strength will permit—say seventy-five or eighty during three months.

Usually, it is more profitable to alter and fatten a boar when three years old than to keep him longer. But, of course, much depends on his value and on our ability to replace him. Fisher Hobbs' celebrated Essex boar, "Emperor," of which we have given a portrait on page 79, was eight years old when his picture was taken. An animal of extraordinary merit may be kept as long as he gets good pigs.

The boar's pen should have a yard attached not less than ten or twelve feet square, and it is better, always, to turn the sows to him, than to turn him out to the sows. If he is sluggish, it is well to have a strong door between this yard and the boar's sleeping pen, so arranged that he can see the sow without being able to get at her until the door is open. Shut the door between the pen and the yard, and then turn the sow into the yard, and let her remain a short time before letting the boar out. The best boar we ever had was exceedingly shy in this matter. He apparently objected to have strangers looking on. We kept him for some years, and by humoring his peculiarities, he proved a very useful animal. He always showed most energy early in the morning, before he had had his breakfast. Some of our neighbors, who had been accustomed to drive their sows to common boars, that would tear down a pen, or push over a fence to get at a sow, were disgusted with the dignified movements of this thorough-bred boar. After waiting and watching a few minutes, they would drive away their sows to some long-nosed, slab-sided brute, while those who exercised a little more patience, were almost invariably rewarded with splendid litters of pigs. The truth of this matter is, that good breeders increase the development of the choice parts of a pig at the expense of the offal; and the ham of a well-bred and well-formed boar has been enlarged at the expense of some portion of the contiguous parts. We have known this carried to such an extreme, that casual observers would suppose they were looking at a barrow-pig. Any one who will contrast a coarse Chester County boar with a refined Essex will understand our meaning.

THE SOW.

The treatment of a sow until she is eight or nine months old does not differ from that of the boar. She should be well fed, and have plenty of exercise. If she is born in March, and is kept growing rapidly, and is of an early maturing breed, she may be allowed to take the boar in November, when about eight months old. She would then have pigs in March, when a year old. This is breeding earlier than is usually recommended, but it must be remembered that we are treating of pigs that have been bred almost exclusively for the purpose of rapid growth while young, and for early maturity. If she is strong and healthy, with good digestive powers, it will not hurt her to have a litter of pigs at a year old, and to have two litters a year afterwards, for two or three years. The breeder, however, must exercise judgment in this matter. It often improves a sow wonderfully to let her get a year or fifteen months old before she takes the boar. And in the case of late fall pigs, we should always be inclined to keep them until the following November before they are served.

The sow, when in pig, should be allowed abundance of food, and as extensive a range as possible until a week or ten days before farrowing. She should then be placed in her pen, and fed with food similar to that which it is intended to give her after she has farrowed. Nothing can be better than skimmed milk and scalded bran, with a little oil-meal, to loosen the bowels, if necessary. Directions for furnishing the pen with litter, etc., have been already given, and need not be repeated here. As, however, a litter of thorough-bred pigs are of considerable value, we would particularly urge that everything be provided in advance that can insure their safety. We have lost a litter of ten pigs that, at ten weeks old, would have brought us two hundred dollars, simply from neglecting to have the pen properly protected beforehand against a severe storm which occurred the night the sow farrowed. Much is said about sows eating their young, but where one pig is lost in this way, a hundred die from damp pens and neglect.

When the sow is shut up by herself in the pen, if she is uneasy, it is well to let her out for an hour or so during the forenoon, letting her in again for her noon meal, and in the course of an hour or so, let her out again, putting her back at feeding time for the night. In this way she will soon become accustomed to the pen. It not unfrequently happens that the sow, at this period, is constipated; and if this is the case, she should be fed on more succulent, and less concentrated, food. We know of nothing better than bran mashes, either alone, or mixed up with linseed tea. If this does not relieve the trouble, give an injection of warm water, with a little soap in it. In obstinate cases, put an ounce of Epsom, or two ounces of Glauber's salt in the injection. This is generally better than giving her medicine, even if she would eat it in her food, which she will seldom do. It is not safe to attempt to drench a sow at this period. A careful attention to the diet, with sufficient exercise, will almost always prevent this trouble.

Our own sows are so quiet that we can do anything with them. And before they farrow, we are in the habit of handling them, rubbing their teats, and getting them thoroughly accustomed to our presence in the pen. If all goes right, it is best to let the sow alone; and, in all cases, it is better to err in giving too little attention or assistance than too much. If the weather is very cold, throw a blanket over the sow; and as soon as a little pig arrives, rub him dry with a little soft straw, and put him to the teats, under the blanket. Be careful, however, not to break the cord too close to the navel, or it may cause blood to flow, and thus weaken the pig. If the sow has been well and properly fed, and is in vigorous condition, the pigs will be strong, and will take hold of the teats in a few minutes. When this is the case, little danger of loss is to be apprehended. If any of the pigs are weak, it often requires considerable care and attention to save them. The great point is to prevent them from becoming chilled and to get them to suck. It is here that the previous petting of the sow and handling of her teats prove useful. You can hold the pig to the teat, and press out some milk with the thumb and finger. It is said that the teats, towards the forelegs, afford the richest milk, and that, as each pig is believed to always keep the teat he first takes possession of, it is well to put the weaker pigs to the forward teats. We cannot speak from experience as to the advantage of this method. In the case of thorough-bred pigs, it will pay to have a man watch the sow the first night, to see that she does not lie on any of the little ones. If the pigs are strong, there will be comparatively little danger after the first night; but we have known a sow to lie on a weak pig and crush it to death when eight or ten days old, and when all danger was supposed to be passed. We once had a sow lie on a sick pig, that was large enough to wean, and hurt it so much that it died in a few hours. If the pigs are strong, it is an easy matter to raise them; but if not, great care will be required. It is, therefore, in all respects, very desirable to have the pigs come strong and healthy, and this is usually the case when the sow and boar are healthy, and are descended from a healthy stock, and when the sow herself is, and always has been, well and properly fed, and has had plenty of fresh air and exercise, with access to charcoal, ashes, and pure water.

When a litter of pigs gets chilled, there is nothing better to revive them than hot chaff from the steam vat, renewed by degrees as often as it gets cool. We have saved pigs in this way that were almost lifeless. Place the chaff along the side of the sow, next to the teats, and put the little pigs on it, and nearly cover them with the chaff, and then throw a blanket over the sow and the pigs and hot chaff. Of course it will be necessary to remain with the sow and watch the pigs; and we have sometimes given them, with advantage, a little warm new milk, or fresh cream, with a teaspoonful of gin or whiskey to three or four tablespoonfuls of the milk or cream. When they revive a little, place them to the teats, and encourage them to suck a little.

In very cold weather, it is often desirable to hang some blankets from the top of the pen around the sow, like the curtains on a tent-bedstead; and by placing several bags of hot chaff inside the curtains, the temperature may be raised several degrees. If more convenient, pails of hot water may be used instead of the hot chaff.

We once had a litter of valuable pigs come one night during a severe cold storm. The kitchen fire was out, and no hot water to be had, but in the steam-house was a barrel of boiled barley. By taking a little from the top it was found to be hot underneath, and we carried six or eight pailfuls of hot barley into the pen, and in this way managed to keep the pigs warm, and save the whole litter.

When the pigs are two weeks old, they will begin to lap a little milk, and a week later, will eat a few oats. The directions already given in a previous chapter are applicable here. When the pigs are a month old, we let the sow out from the pigs in the morning, after breakfast, and again after dinner, feeding the pigs while the sow is away. At first, the sow is kept out only an hour or so at a time, and as the pigs get older, she may be kept out longer. In this way the little pigs will eat more food, and will not draw so much on the strength of the sow. In the case of thorough-bred pigs, where it is desirable to have the sows breed as long as they produce strong litters of ten or a dozen, this is quite an important point. Great care must be taken not to tax the strength of the sow too much. Little pigs, of a good breed, grow so rapidly, that they require much more food than ordinary pigs, while the sow has been so refined by breeding, that she is seldom strong enough to stand the drain, when the pigs depend entirely on her for food.

The pigs will do better to remain with the sow until they are two months old; and if they are well fed, and are gradually weaned in the way above recommended, the sow will suffer no harm.

According to the experiments of Doctor Miles, previously alluded to, Essex pigs, about three weeks old, ate $3^1|_2$ lbs. of new milk, each, per day. The next week they ate nearly 7 lbs. of milk, each, per day. From this, it appears that a litter of ten pigs, a month or five weeks old, will eat over 30 quarts of new milk a day, or more than is ordinarily given by the best cows. We present these facts here to show what an immense drain a suckling sow is called upon to sustain. We have often observed how rapidly such a sow loses flesh after the third week. No matter how fat she may have been, and how much of the richest food she is allowed, she will soon get very thin unless the pigs are induced to eat other food than that which the mother supplies.

The milk of the sow is richer than that of any other domestic animal. Milk is derived from the blood, and this is derived either directly from the food, or from the flesh and fat stored up in the animal. It is, therefore, easy to understand that, when a sow is called upon to give as much milk as one of the largest and best cows, it must tax her digestive powers to the utmost, or rapidly convert her flesh into blood and milk.

We are particularly anxious to call attention to this matter, as we deem it one of the most important points in the management of thorough-bred pigs. A litter of ten pigs, at birth, weighs about 15 lbs., and at six weeks old, sometimes as much as 250 lbs., or nearly or quite as much as the mother herself weighs, in many cases. It is evident that this enormous growth must require a large amount of food from somewhere. From whence is it obtained? In thorough-bred pigs, we must have as rapid growth as possible while young, or the breed deteriorates. The offspring of pigs whose growth is checked while young from want of food, will, in some degree, lose the capacity of growth, even though abundance of food is furnished. The sins of the owners of the parents are visited on the owners of the children. The pigs have been bred for the very purpose of growing rapidly, and they cannot grow without food. To expect a thorough-bred sow (refined down to the last degree,) to raise a litter of pigs (inheriting a tendency to rapid growth), with no more food than a common sow with a litter of common pigs, is unreasonable. The thorough-bred sow and pigs require, and must have, better food, and more attention than the common pigs.

A first-class thorough-bred sow, that produces eight or ten pigs at the first litter, and proves a good mother and nurse, is a very valuable animal, and it will pay well to take care of her. For the first two weeks after farrowing, little change will apparently take place in her condition. The scales would doubtless show that she has lost weight, but it is from the inside fat, which finds its way into the milk for the nourishment of the young. All animals lay up fat for this purpose, and it is not necessary to furnish a large quantity of rich food for the sow for the first week after farrowing. She should have all the cooling drinks she requires, and food that is easily digested, such as milk and bran mashes, and later, oat-meal or barley-meal. After the second week, give richer food, but be careful that it is not rich enough to derange the stomach of the sow, and produce diarrhœa in the little pigs. Boiled barley, given in connection with the milk and bran, is excellent. Let it be thoroughly boiled. Soak it in water for twelve hours, and afterwards boil it in the same water until it bursts open. Three weeks after farrowing is the critical time for the sow. The pigs begin to require much more milk, and are constantly pulling at her. She will begin to fall off in flesh, and this is not, in itself, objectionable, provided it is not carried too far. It is here that the breeder must exercise his best judgment. The sow must have a liberal and regular supply of nutritious food. But be very careful not to give her a comparatively innutritious food one day, and a full supply of rich food the next. The true plan, as we have before said, but it cannot be too often repeated, is to *feed the little pigs*, and thus lessen their demands on the mother. Give them a little new milk from the cow, and take pains to teach them to drink it. If you teach one to drink, the others will be likely to follow his example. A little sugar or molasses in the milk will prove acceptable to the pigs. In a few days, mix a little scalded or boiled oat-meal with the milk, and gradually increase the quantity as their appetites increase. A little boiled barley may also be given, and throw them a handful of whole oats on the floor of their pen, for them to crack and exercise their teeth on. In this way you can save the strength of the sow, and we deem this one of the most important points in breeding, especially with the first litter.

In the natural state, sows do not have more than half as many pigs at a litter as the improved breeds, and they do not grow half as fast, and consequently do not require more than half as much milk. Those who talk so much about following " Nature," seem to forget these facts. Our object is to improve on nature, and to do this we have to provide improved conditions. A thorough-bred pig is a work of art, and its production calls for intelligence, thought, care, patience, and perseverance.

We once had two valuable thorough-bred sows that farrowed their first litters in February. They had ten pigs each. Through carelessness, one whole litter was frozen to death. We took a couple of the pigs from the other litter, and gave them to the sow that had lost her litter, and these also died. The other sow raised the eight pigs, and they did well. The sow was left in charge of an ordinary man, and by the time the pigs were five weeks old, she was as thin as a rail. The pigs were not weaned until nine weeks old. She nourished them at the expense of her own flesh, and, as it turned out, at the expense of her strength also. She did not recover from the effects of the drain on her constitution for six months, and did not take the boar again until the following October. In the meantime, the sow which lost all her pigs took the boar in two weeks, and had a litter of ten pigs in July, worth, at two months old, $20 each. We mention this fact to show that it will pay to take particular care of young sows, and to guard against overtaxing their strength and constitution. We must do this, not only by giving the sow the best of care and proper food, but also by feeding the little pigs, and doing all that we can to prevent the sow from giving them too much milk after they are three or four weeks old.

A sow will often take the boar in three or four days after farrowing. In the case of large, coarse, common sows, this is sometimes desirable, but rarely in the case of thorough-breds. It is better to wait until the pigs are weaned. If the little pigs have been fed as above recommended, so that they have not taxed too much the strength of the sow, she will often take the boar in a few days, or, at farthest, in two or three weeks. She should have plenty of nutritious food and moderate exercise for the first month after the pigs are weaned. After this, she should have all the food she can eat, but should, if possible, be compelled to take some exercise in order to get it.

MANAGEMENT OF THE YOUNG PIGS.

The pigs, as before said, should be gradually weaned. They do better to remain with the sow until eight or ten weeks old, but we would commence weaning them when three weeks old. Let out the sow from them—at first, for an hour or so at a time, gradually extending the time as they get older. When a month old, they may be allowed to go out with the sow for an hour or two in mild weather, but not while the sun is very hot, as, in some breeds, our hot sun will blister the backs of young pigs. When five weeks old, they may go out into the pasture while the sow is kept in the pen. The little pigs need more exercise at this time than the mother. The secretion of milk, in her case, is equivalent to a considerable amount of exercise, and she should not be *obliged* to take exercise in order to get food.

The most common complaints of little pigs are diarrhœa and colds. The former is caused by giving the sow improper food, or a too sudden change of diet, or by irregular feeding, or from want of pure water and fresh air. We once had a few cooked beans that had been left in the steam-barrel until they decomposed. They were thrown on to the manure heap, and a sow, which was suckling pigs, ate some of them. Two days afterwards, the whole litter was seized with violent diarrhœa, and one of them died in the course of two or three days. It was the worst case of the kind we ever had, and the diarrhœa continued for four or five days, and was not stopped until we gave the pigs two or three drops of laudanum each, at night, in some fresh cream, with a teaspoon, and repeated the dose the next morning. This effected a cure, but the pigs did not regain their thrifty growth for a week or ten days. We should add that the sow continued perfectly well, and manifested no symptoms of the complaint. As a general rule, no medicine will be required. Change the food of the mother, and let her go out into the air, but let the little pigs remain in the pen, and see that they are warm and comfortable. The less they are disturbed, and the more they sleep, the sooner will they recover. It is also very important to keep the pen clean and well ventilated. Nothing can be worse than to leave the evacuations in the pen. Scatter some dry earth about the pen to absorb the offensive gases. Let the feeding apartment also be dusted over with dry earth or soil of any kind that can be obtained, and then scraped, and swept, and washed, and a little dry straw, or chaff, or sawdust, be spread on it, to prevent dampness. Scald the pig troughs with *boiling* water, and make them sweet and clean. Let this be done every day. The attendant should understand that the scours are an evidence of negligence or carelessness.

The same may be said of coughs or colds. Damp pens, exposure to a cold storm, too much litter at one time, and too little at another, or suffering it to remain until it gets damp, are the chief causes of colds, with all their attendant disorders. An ounce of Epsom salt, given to the sow in her food, twice a day, will be beneficial to the little pigs. But it is not often that pigs are affected with colds until after they are weaned, and in this case a few salts, either Epsom, Glauber's, or Rochelle, as most convenient, may be given in the food—say a teaspoonful of Epsom or Rochelle salts, to a three-months-old pig, or a tablespoonful of Glauber's salt, given in the food twice a day, with a little gentian or ginger, or some other tonic. Fresh air is very important, and in mild weather they should be allowed to run in the pasture, but should be permitted to return to their pen whenever they wish. Let the pen be made as dry and comfortable as possible;

give succulent food, and guard against constipation, and in a few days the pigs will be better. In our own experience, we have never happened to have any serious trouble from this cause, but we once sent a pair of valuable pigs to a gentleman in Illinois, and the boar, nine or ten weeks old, and a very strong and apparently healthy one, caught cold on the route, and though he received good care, died in a week or so afterwards.

The great point in the management of young pigs is, to keep them growing rapidly. If strong and vigorous, they are seldom liable to any disease, and if attacked, soon throw it off. We think it advantageous to pet them and make them as tame as possible. They are fond of being rubbed with a brush, and have not the slightest objection to a good Irish scratching, especially in the holes and corners about the head, where they cannot scratch themselves without unusual exertion. We are in the habit of taking hold of our young pigs back of the ears, and when they get used to it, they regard it as indicating a desire for a frolic. If well fed, well petted, and in high health, they enjoy a frolic as much as a pair of young dogs. At three months old, the boar pigs should be separated from the sows.

CHAPTER XXIII.

THE PROFIT OF RAISING THOROUGH-BRED PIGS.

A farmer who reads the preceding chapter will be very apt to ask—" Will it pay to be at all this trouble to raise pigs? Will it not be better to keep a kind that does not require so much attention?"

In the first place, it should be remarked that, we do not advocate keeping thorough-bred pigs to be fatted and sold to the butcher. They are raised for the purpose of improving our ordinary stock; and we have already attempted to show what is the value of a thorough-bred boar for this purpose. Suffice it to say here, that he is worth much more than he is ordinarily sold for. We believe that reliable breeders of thorough-bred pigs are often unable to supply the demand for boars; and it is certain that, as their value for improving our common pigs becomes more generally recognized, the demand will become far greater. At the present time, not one boar in a thousand, kept for use in the country, is thorough-bred. The American agricultural press, which is becoming a mighty power for good in the land, is doing valuable service in calling attention to the importance of using none but thorough-bred males of all kinds of stock, and the prospects of breeders never were more encouraging than now. As general intelligence and civilization increase, so increases the demand for flesh meat of good quality; and the prices paid for it warrant us in using every means in our power for increasing the supply. In the future, as in the past, the price of pork will fluctuate; but with our facilities for transportation, and the ease with which pork can be cured and shipped to any part of the world, the American farmer is pretty certain of getting a fair price for his pigs. The English farmers are enabled to compete with the pork made from our cheap corn-growing sections by paying more attention to the improved breeds, and by furnishing a superior article. The American farmers of the Eastern and Middle States must do the same thing in order to successfully compete with raisers of pork in the cheap corn-growing sections of the West; and the first step is to introduce thorough-bred boars of the best breeds.

As long as breeders can sell their pigs at $20 each when two months old, it will pay to bestow a good deal of attention on their management. An English breeder is said to have made enough out of his pigs to "build a church." Many American breeders of Chester County and the Jefferson County pigs have **made a great deal of money by the business.**

CHAPTER XXIV.

COOKING FOOD FOR PIGS.

Nearly all farmers cook more or less food for their pigs. Comparatively few do it systematically and regularly throughout the year. Potatoes, pumpkins, and food of this class, is almost invariably cooked in this country, the general plan being to boil or steam the potatoes or pumpkins, and after they are cooked, mash them up with meal, either in the vessel in which they are cooked, or in the feed tub. If the meal is mixed with the cooked food while it is boiling hot, and the mass is then covered carefully for a few hours, to retain the heat, the meal becomes soft, and is, in fact, more or less cooked, according to the skill and judgment with which the operation is performed.

In England, Swede turnips are often cooked in this way, and mixed with barley or Indian corn-meal. But they are considered far inferior to potatoes as food for pigs. Of late years, the turnips, potatoes, etc., are ground, or crushed, and the pulp, as it comes from the machine, is mixed with meal. This mixture of meal and pulped roots is sometimes steamed, but it is more generally fed without cooking, being simply allowed to remain in a heap until it becomes warm from fermentation. In this way the particles of meal are softened and broken up, and are supposed to be more readily digested by the animals. As to whether it is more economical to feed raw potatoes with raw meal or grain, or to cook them, there seems to be no question. We have never known any one who has tried steaming or boiling, with even ordinary conveniences, that was not perfectly satisfied that it was more profitable than to feed raw. We may assume that this fact is established by common experience. But, on the other hand, as between cooking and pulping, the question may be considered an open one—that is to say, as to whether it is more economical to steam roots and meal, or to pulp the roots and mix meal with the pulp, and then allow the mixture to ferment, has not been satisfactorily determined. It depends, probably, a good deal on the conveniences for doing the work.

If we might hazard an opinion, from a quite limited experience, we should say that, for store pigs and breeding stock, we should prefer, where there are good conveniences for steaming, to pulp the roots, mix them with sufficient hay chaff to absorb the juice, and then add a little meal, and steam the whole mixture together. The clover hay imparts an agreeable flavor to the cooked mass, and the pigs eat it with far more avidity than they will eat the raw pulp and meal mixture. If we can winter our pigs on roots, and clover hay, with a little meal, one of the chief objections to keeping a large stock of pigs is entirely removed. They are then kept on food, the production of which enriches, rather than impoverishes, the soil, while the manure from it is of the richest and most valuable description.

Where pigs are kept for the purpose of supplying the demand for choice fresh pork, cooking will probably be found essential to success. The pigs should be ready for market at from four to five months old. In proportion to the food consumed, young pigs (and probably all other animals) grow much more rapidly than older ones. But if they are to grow rapidly, and fatten at the same time, they must have the richest and most easily digestible food. Of course they must be fed with judgment, varying the food as occasion requires, and sometimes giving raw grain, but our main dependence must be steamed roots and meal; or, in the absence of roots, we must have cooked meal, with sufficient steamed hay or grass to fill the stomach, and keep the bowels regular. The richer the food, provided the pigs can eat enough of it to fill their stomachs three times a day, without producing constipation or scours, the more rapidly will they fatten.

There is a sense in which it may truly be said, that cooking adds nothing to the amount of nutriment in food. All that can be claimed for it is that it increases the *digestibility* of the food. To what extent this takes place has not been determined. In fact, the whole subject is surrounded with difficulty.

In Chapter III. we have endeavored to show how important it is to obtain animals that will eat and digest a large amount of food. And it may be recollected that, in Dr. Miles' experiments (see page 122), 100 lbs. of meal, eaten by *one* pig, gave an increase of $19^1/_2$ lbs., while the same quantity, eaten by *two* pigs, gave only an increase of 3 lbs. The food was of the same character, and the difference in the results is due to the better appetite and digestive powers of the pig that ate double the amount of food. But the fact shows how important it is to provide food that pigs will eat and digest.

Those who advocate cooking food for animals, frequently assert that it "saves one-quarter of the food." We know of no satisfactory experiments which establish the fact. And, at any rate, it may safely be asserted that the saving of food is only a very small part of the advantage to be gained from cooking. What we should aim at in breeding and feeding, is to get pigs to eat 25 per cent *more*, rather than 25 per cent *less*, food. We have assumed (see page 22) that 75 per cent of the food a pig eats is ordinarily required to support the vital functions. If a pig eats 100 lbs. of corn in a month, and gains 20 lbs., we assume that 75 lbs. are used to support the vital functions, and 25 lbs. are left available for growth. On this supposition, take three pigs, and put them in separate pens. Feed one whole raw corn, another raw corn-meal, and another cooked corn-meal, and assume that one eats

87½ lbs. during the month, the other 100 lbs., and the other 125 lbs., and we may then get the following results:

	Food consumed.	Food required to sustain the vital functions.	Food available for increase of growth.	Growth of Pigs.
No. 1, Whole Corn, Raw	87½ lbs.	75 lbs.	12½ lbs.	10 lbs.
No. 2, Raw Meal	100 "	75 "	25 "	20 "
No. 3, Cooked Meal	125 "	75 "	50 "	40 "

This is assuming that the grinding and cooking do not add anything to the intrinsic nutriment of the food, but merely render it more digestible. We assume that when whole raw corn is fed, the pig can only digest 87½ lbs. per month, but when ground and cooked, it can digest 125 lbs., and gains *four* times as fast. Of course these figures are only hypothetical. They may, or may not, be true. We give them merely to illustrate our meaning, and to show how important it is to have pigs that can eat and digest a large amount of food—and consequently how important it is to provide food readily digestible.

It may be true that cooking enables the pigs to fatten on less food, but if so, it must be owing to the inability of the pigs to digest the raw food. They must void a portion of it in an undigested state. To a certain extent this can be avoided by feeding less grain, and furnishing the necessary bulk to fill the stomach by supplying a portion of less concentrated food. When pigs are allowed the run of a clover pasture, they may be fed whole grain without much loss from passing it in an undigested state. The feeder, by examining the feces, can tell how much grain he can feed without loss. If he feeds more than the pigs can digest, he suffers a loss of grain; but if he feeds less, he suffers a certain amount of digestive power to run to waste. His profits will depend very much on his ability to guard against loss from either source.

We cannot too often call attention to the great mistake which many farmers make in not-feeding any grain to their pigs during the summer and autumn, while at pasture. It is not uncommon to furnish the pigs nothing but grass and the slops from the house until the time the corn crop is ready to husk. They are then shut up in a pen, and thrown whole corn on the ear. The pigs have been accustomed to a bulky food, from which they can extract little more than sufficient nutriment to keep them alive, when, suddenly, they are shut up, and allowed nothing but food containing, in a given bulk, three or four times as much nutriment. What wonder if a portion of it is voided in an undigested state? If the pig fills his stomach, what else can he do with it? His powers of digestion and assimilation are not three times as great to-day as they were yesterday, when he had nothing but grass. How much more reasonable it would be to feed him a little corn when at grass, and a little grass, or other succulent food, when shut up to fatten!

The corn fed to a pig while at grass increases his powers of digestion and assimilation, and as he approaches maturity, he will be able to digest and assimilate more concentrated food. The aim must be to furnish him all he can possibly eat, digest, and assimilate. It is here that cooking comes to our aid. It enables us to "crowd" the fattening pigs forward rapidly to maturity. It is a costly process, feeding pigs wholly on grain, and we must shorten the time as much as possible. The pigs should be kept growing rapidly during the summer, increasing the supply of grain as the pigs get older, and when shut up to fatten, four or five weeks feeding on rich, cooked food, should fill them up with lard.

By looking at our market reports, it will be seen that there is a difference of two or three cents per pound in the price of pigs, according to their condition and quality. And, in point of fact, there is even a still greater difference in the intrinsic value of the pork and lard to the consumers.

This is a point that should not be overlooked in estimating the advantages of liberal feeding. Take two litters of ten pigs, each born, say, the first of September. Let both litters have the run of a barn-yard, with the slops from the house, dairy, etc. Let one litter have nothing but what they can pick up. Let the other have what they can pick up, and be supplied with a feed of grain, *in addition*, that shall send them to bed every night with a full stomach. By the first of May, the one litter should weigh 200 lbs. each; the other would be better than the average if they weigh 100 lbs. each. Then let both litters have the run of a pasture, with the slops from the house, etc. Let the one have nothing else, and the other be allowed a little grain every day—enough to fill their stomachs every night, and make them sleep comfortably. By the first of October, the one litter will weigh—say 350 lbs., the other 150 lbs. each. Then shut them both up to fatten. Let both litters have all the corn they can eat. Give one cooked corn-meal, and the other corn in the ear. In a month, the one should weigh 400 lbs. each, the other 175 lbs. each. Last year the one litter would have sold, say for 10 cents per lb., live weight, the other for $7\frac{1}{2}$ cents, and we have the following results:

10 pigs, 400 lbs. each, at 10 cents.................................... $400.00
10 pigs, 175 lbs. each, at 7½ cents................................... 131.25

To pay for extra feed.. $268.75

We may estimate the extra feed as equal to an average of half a pint of corn per day, each, from the first of October (when the pigs are a month old) to the first of December, say half a bushel of corn for each pig. From the first of December to the first of May, say one pint per day, or less than $2\frac{1}{2}$ bushels for each pig. From May until October, allow one quart per day, or, say 5 bushels to each pig. This would be 8 bushels of corn to each pig. And we have no sort of doubt that, in the circumstances assumed, this 8 bushels of extra corn on each pig, or 80 bushels in all, would make the difference shown by the figures just given.

To cook grain for pigs merely for the sake of "making it go further," will seldom pay on ordinary farms. This is particularly the case where grain is comparatively cheap, and fuel dear. It is profitable only when adopted for the purpose of enabling the pigs to eat and digest a greater quantity of food, and bring them rapidly forward for market.

And it is still an open question whether we cannot adopt some cheaper method of increasing the digestibility of grain than grinding or cooking it. Where grain can be ground cheaply on the farm, we would grind or crush it for all kinds of stock. But when it has to be sent some distance to a mill, it is worth while to see if we cannot prepare it at home.

In Mr. Lawes' experiments on sheep, eight Hampshire Down sheep were put in two pens, four in each pen, and allowed all the mangel wurzel they would eat. Pen 1 was allowed 1 lb. of barley for each sheep, per day, the barley being coarsely ground. Pen 2 was allowed the same quantity of barley, also coarsely ground, but before being fed, it was *soaked in cold water for 24 or 36 hours*. The experiment lasted ten weeks. The following are the results:

	FOOD CONSUMED BY EACH SHEEP PER WEEK.		*Increase of each sheep per week.*
	Barley.	Mangels.	
Pen 1—Barley-meal, fed *dry*........	7 lbs.	96½ lbs.	2 lbs., ½ oz.
Pen 2— " " " *soaked*.....	7 "	117½ "	2 " 8½ "

Soaking the barley enabled the sheep to eat more food, and grow 25 per cent faster than those having dry barley. Had the sheep been allowed *more* of the soaked barley, the result would probably have been still more in favor of the practice. One of the sheep in pen 2 gained 4 lbs. per week. He probably got more than his just proportion of barley, and the other three were obliged to make up the deficiency in eating more mangels. And so the total gain, in proportion to total food consumed, is not as great as it otherwise would have been. The amount, of actual dry matter in the food, required to produce 1 lb. of increase, is nearly identical in both pens—$8^1|_2$ lbs.

With pigs, when they are allowed all the grain they will eat, we have no doubt that soaking the grain would show still better results. In this country, where we feed so few roots, the experience of farmers indicates that they have a greater nutritive value than the mere amount of nutriment they contain would indicate. This is attributable, to a certain extent, to the fact that the food in the roots is intimately mixed with a large amount of water. Now, by soaking grain, it absorbs a considerable amount of water. Barley will absorb nearly half its weight of cold water. When cooked until it bursts open, it doubtless absorbs a still greater quantity. In the absence of roots, therefore, we may obtain food somewhat resembling them by soaking or cooking grain. With the requisite number of tubs, it is an easy matter to have a constant supply of soaked grain for pigs or other stock. In fact, it would not be a difficult matter to soak the grain until it had absorbed all the water it would take up, and then keep it in a mass, from twelve to sixteen inches deep, until it begins to sprout, whereby a portion of the starch is converted into sugar. As the grain grows, it must be spread out in a thinner layer. But it is probably better to feed it out soon after it commences to sprout, as the process of germination is attended with more or less loss of carbon.

Where whole grain is steamed, there is a great saving of time and fuel by soaking the grain for 24 or 36 hours before letting on the steam. We are inclined to think that it can be cooked in this way to fully as much advantage as when it is ground into meal. Grain, whether whole, or ground into meal, cannot be steamed without water, and if it could be, it is doubtful if it would be as good for the animals. The absorption of the water, and having it intimately mixed with the meal, is one of the advantages of cooking. Boussingault well says: "The absolute necessity of a sufficient degree of moistness in the food, in order to secure its due and easy digestion, greatly countenances the practice which is beginning to be introduced in some places of steeping hay for some time in water before giving it to cattle." We think there can be no question that soaking or cooking food renders it much more easily digestible, and if so, the advantages of the practice, where liberal feeding is adopted, cannot be doubted.

We may add that whole grain, thoroughly soaked or boiled, swells to about double its bulk, and consequently, in feeding, we should allow, *at least*, twice the quantity that the pigs eat when dry. To attain the best results, we should watch the pigs eating, and when they have eaten up all clean, give a little more, and encourage them to eat as much as possible. There is an amusing story in the *American Agriculturist* that illustrates the importance of inducing pigs to eat as much as possible.

"A good story was lately told us of several neighbors who, year after year, vied with one another in trying to produce the fattest hog, each taking a pig from the same litter, or in some way starting fair, and square with pigs of the same age and size, and doing his best to make it as fat as possible before Christmas. One of the farmers invariably beat the others out and out so thoroughly, that his good luck could never be accounted for as accidental. The secret he kept to himself, but being watched by some one determined to find it out, the discovery was made that jealousy is a grand appetizer for hogs. First the pet monster was allowed to fill himself to his heart's content, and when his appetite was satiated, a half-starved shoat was let in to the pen by a side door. The fat one would at once begin to fight it off, and meanwhile, to gorge himself, simply to prevent the poor, squealing victim of unsatisfied cravings getting any food. This was a daily

Fig. 53.—JEALOUSY AN AID IN FATTENING.

programme, and the result was as stated. The fact is worth bearing in mind for, in preparing hogs for exhibition, or for some reason, we are often desirous of expediting the fattening process."

CHAPTER XXV.

SUMMARY.

It may be well, in conclusion, to state a few facts that may have been given in previous chapters, but which it may be convenient to place here in a concise form for reference.

The leading breeds of English pigs are Berkshire, Essex, and Yorkshire. The Essex are entirely *black*, the Berkshire are also dark colored pigs, but not so black as the Essex, and have also white spots on the head and feet. There are large and small Berkshires. The Yorkshires are white, but occasionally dark spots show themselves on the skin, and these are not considered decisive evidence that the pigs are not thorough-bred. There are small, medium, and large, or mammoth, Yorkshires.

The Essex will, at maturity, dress from 400 to 450 lbs. They are the largest of the small breeds. Berkshires often exceed this weight, but when such is the case, they would be classed as Large Berkshires. The Prince Albert Suffolks are small Yorkshires.

The leading breeds, originating in the United States, are the Cheshires, or Jefferson County, the Chester Whites, or Chester County, and the Magie, or Butler County pigs. The China-Polands, or China, and Big Polands, are said to be the same breed as the Magie, or Butler County. The Illinois Swine Breeders' Association, at its meeting in 1870, resolved to call them the "Magie" breed. They are a large, coarse breed, with black and white, and occasionally sandy, spots. Like the Chester Whites, they will doubtless afford splendid sows for crossing with Essex, Berkshire, or other refined thorough-

SUMMARY.

bred boars. The Jefferson County are a very handsome white breed, essentially Yorkshires.

Pigs should always have access to fresh water. No matter how "sloppy" the food is, or how much dish-water is furnished, they should be furnished with pure water. We are satisfied that pigs often suffer for want of it.

Salt, sulphur, charcoal, ashes, bone-dust, or superphosphate, should occasionally be placed where the pigs can eat what they wish of them.

If *thoroughly* boiled, pigs will eat beans, though they are not fond of them. Peas they eat with avidity, and when as cheap as corn, should be fed in preference, as they afford much the richest manure. Half peas and half corn is probably better than either alone. Peas make very firm pork.

Oil-cake, when fed in large quantities, injures the flavor and quality of the pork, but we have fed small quantities of it, with decided advantage to the health and rapid growth of the pigs, without any apparent injury to the lard or pork. It is quite useful for breeding sows. It keeps the bowels loose, and increases the quantity and quality of the milk.

Bran, except in small quantities, is not a valuable food for fattening pigs. It is too bulky. But when rich, concentrated food is given, such as corn, barley, peas, or oil-cake, pigs should be allowed all the bran they will eat, placed in a separate trough. In this way it becomes a very useful and almost indispensable article to the pig feeder. It is also very useful for breeding sows.

The best roots to raise for pigs are parsnips and mangel wurzel.

The period of gestation in a sow is almost invariably sixteen weeks. In three or four days after pigging, a sow in good condition will generally take the boar. But, as a rule, it is not well to allow it. If she passes this period, she will not take the boar until after the pigs are weaned.

If she fails the first time, she will "come round again" in from two to three weeks.

For mild cases of diarrhœa, nothing is better than fresh, skimmed milk thickened with wheat flour.

Pigs should be castrated a week or so before they are weaned.

Nothing in the management of pigs is more important than to provide a trough for the sucking pigs, separate from the sow, and to commence feeding them when two or three weeks old.

Many of the diseases of pigs are contagious, and the instant a pig is observed to be sick, it should be removed to a separate pen. And it would be well to regard this single case of sickness as an indication that something is wrong in the general management of the pigs. Clean out the pens, scald the troughs, scrape out all decaying matter from under and around them, sprinkle chloride of lime about the pen, or, what is probably better, carbolic acid. Dry earth is a cheap and excellent disinfectant. Use it liberally at all times. Whitewash the walls of the pens. Wash all the inside and outside wood-work, troughs, plank floors, etc., with crude petroleum. It is the cheapest and best antiseptic yet discovered.

To destroy lice, wash the pig all over with crude petroleum, and the next day give him a thorough washing with warm soft water and soap, with the free use of a scrubbing brush.

In the absence of anything better, we use petroleum for all diseases of the skin in pigs, flesh wounds, etc.

For a mild blister, in cases of cold, or threatened inflammation of the lungs, foment the body, under the forelegs, for an hour or so with cloths wrung out of hot water, and rub on a little saleratus or soda occasionally during the operation, to soften the skin, then apply petroleum. This will then act as a mild irritant, and heal at the same time.

Mange, or itch, is caused by a minute insect, which is probably hatched from eggs adhering to the skin. There is no way of curing it, or of preventing its spread, except by killing the insects and their eggs—not only on the pigs themselves, but also on the sides of the pens, posts, or anything that the diseased pig rubs against. To destroy them on the wood-work, nothing is probably so good as petroleum, and though we have not tried it, we have little doubt that it would also cure the pigs, especially if applied before the disease has made much headway.

The disease usually manifests itself on the thin skin under the armpits and thighs, and inside the forelegs. At first, small red blotches or pimples appear, and these gradually spread as the insects multiply and burrow under the skin. It is well to give sulphur and other cooling medicine in the food, but the real aim must be to kill the insects by the prompt and continued use of carbolic acid, petroleum, or a strong decoction of tobacco. Solutions of arsenic and corrosive sublimate are used in severe cases, but are dangerous articles to place in the hands of inexperienced persons. " Unguentum," or mercurial ointment, is efficacious, but is not easily applied.

Measles should be regarded as an evidence of bad treatment. In-and-in breeding, dirty pens, impure food, and especially allowing them to eat the droppings of other animals, are probably some of the causes of this disease. Where fattening pigs are fed on whole corn, and the store pigs or breeding sows are allowed to eat their droppings, which they frequently do, it should surprise no one if these pigs, or, still more likely, their offspring, are attacked with measles. From the investigations of Dr. Thudicum and others, it is now clearly proved that measles in pigs is caused by small entozoa, or internal parasites, which are embryo forms of the common tapeworm. Measly pork is a fruitful source of tapeworms, and is unfit for human food. We cannot too earnestly caution our readers against breeding from pigs that have ever been affected with measles, or allowing their breeding sows to eat the droppings of other animals, and especially of their own. Raw flesh meat, too, should never be fed to pigs. It contains the embryo tapeworms, and will be quite likely to produce measles either in the pigs eating it or in their offspring. Dogs are notoriously subject to tapeworms (probably from eating raw flesh), and where the dog tax is not enforced, we may expect measly pork.

The seat of measles is the cellular matter immediately under the skin. On the skin itself, in pigs affected with this disease, will be found a number of small watery pustules, of a reddish color, and it is attended with cough, fever, pustules under the tongue, discharge from the nose, running from the eyes, weakness of the hind legs, and other indications of general debility. Unless neglected, the disease seldom proves fatal. Sulphur, saltpeter, Epsom salts, or other cooling medicines should be given, with a liberal supply of wholesome, nutritious, and easily digested food.

Rheumatism is not an uncommon disease, especially in thorough-bred pigs, when kept in damp sties, or furnished with rich food one week and poor food the next, or kept in a warm, ill-ventilated sty, and then exposed to storms, and otherwise badly treated. The remedy is Rochelle salts, good treatment, and liberal feeding. Give the salt for two or three days, say one ounce a day for a 100-lb. pig, and less, or more, according to size, and then omit them for a few days.

Protrusion of the rectum, especially with young pigs suffering from a severe attack of diarrhœa, is not uncommon. Wash the gut with warm water, rub on a little laudanum, and then gently press the part back into its place, pushing up the finger for a short distance. A little sucking pig may have five drops of laudanum.

Pigs should be provided with scratching posts, having auger-holes for pegs at different heights, to accommodate pigs of different sizes.

Stephens, in his "Book of the Farm," gives the following description of what may be considered the perfection of form in a fat pig: "The back should be nearly straight, and though arched a little from head to tail, that is no fault. The back should be uniformly broad, and rounded across along the whole body. The touch all along the back should be firm, but springy, the thinnest skin springing most. The shoulders, sides, and hams, should be deep perpendicularly, and in a straight line from shoulder to ham. The closing behind should be filled up; the legs short and bone small; the neck short, thick, and deep; the cheeks round, and filled out; the face straight, nose fine, eyes bright, ears pricked, and the head small in proportion to the body. A curled tail is indicative of a strong back. All these characteristics," he adds, "may be seen in the figure of the brood sow (fig. 52, page 180), though, of course, the sow is not in the fattened state."

CHAPTER XXVI.

APPENDIX.

J. Mackelcan, Esq., one of the editors of the Canada Farmer, and an intelligent and careful observer, favors us with the following notes in regard to his management of pigs.

"My plan of keeping store hogs over winter was to give them a good warm sty, with abundance of room and well littered with straw. They were fed on a mess made of boiled Swede turnips, mixed with pea chaff and finely cut clover hay; the turnips, after being boiled soft, were placed in a barrel, and the chaff and cut hay mashed into them. In addition, they got all the refuse of the kitchen; what milk that could be spared from the dairy being given to the late dropped fall pigs, which had a separate sty to themselves. As soon as the clover was well up in spring, they had the run of a clover field, on which they seemed to thrive, so that, when put up to fatten, at 12 to 16 months old they were about 300 lbs. weight each. Being in good condition, the process of fatting did not take more than three or four weeks; they were allowed all they could eat of peas that had been soaked in water until they were soft and had begun to ferment. Generally speaking, hogs are fatted here by simply giving them hard, whole peas as much as they can eat for about a month, sometimes in the field where they grew, the hogs being put up in a corner and fed from the stack; but it is a wasteful process. The best farmers prefer to either grind the peas and then mix with a little water, enough to make into dough, or, if there is no mill near enough to grind them, to soak the whole peas in water until soft, and then feed to the hogs. The Berkshires (the breed I kept) seem to have an aptitude for eating and thriving on clover; my plan with the young spring pigs was to take them from the sow at eight weeks old, shut them up for a few days, and feed on sour milk or buttermilk in which a little shorts or meal had been stirred. As soon as the clover was pretty well grown, say about the beginning of May, put them in a small paddock by themselves. The paddock must be well seeded with a succulent growth of young clover, and can be made of rails or boards in a corner of a clover field, but must be close enough near the bottom to keep the pigs from getting out. To prevent rooting, they had better be ringed. The young pigs will live and thrive on the clover all summer as long as there is plenty of it. In addition, they should have all the spare milk or whey from the dairy, with some meal occasionally, or, if there is no milk, allow each half a pound of meal per day in water. They must have enough to drink, a little salt once in a while, a shed with a tight roof to shelter them from rain storms or hot suns, and a few shovelfuls of dry ashes in which to wallow and keep off lice. This last may be omitted or only given once in a while. For young pigs, meal should always be cooked or scalded, as raw meal is apt to give them the scours. They should also have free access to charcoal. It is not good for them to eat ashes, nor will they, if they have charcoal; but an ash heap to wallow in will keep them free from lice and fleas. I should also add that my store hogs readily eat fresh cut, green clover, so that, if they have but a small paddock and eat it all down, they can be fed cut clover thrown over the fence to them."

F. W. Stone, Esq., Moreton Lodge, Guelph, Ontario, writes:

"I consider the improved Berkshire the most useful breed for farmers. With pigs, as with every other kind of improved stock, farmers should use nothing but pure-bred male animals. Many farmers send their sows to a pure-bred boar, and are so well pleased with the young pigs, that they select one of them for a boar, and in this way the improvement is soon lost. * * There are many unprincipled men who sell grades for pure breeds, and those who purchase them are disappointed in trying to improve their stock. The breeders of pure-bred stock suffer more from the false representations of such persons than in any other way. Parties, when commencing to breed, or wishing to improve their common stock, should purchase only from reliable breeders, and not from jobbers or traders, who sell anything they can make money by. The young breeder should select the most perfect animals he can find. It is better, in commencing, to invest money in quality rather than in numbers.

"I believe it is better for young sows not to have pigs until they are 14 or 16 months old, though, if the pigs have been well fed, and properly cared for since they were farrowed, good litters may often be obtained at 12 months. A sow, not well fed, is generally pulled down too much to gain the size she otherwise would, by having her first litter before she is 12 months old.

"In Canada, pigs are generally fed with pea-meal, or peas and oats, chopped and mixed with potatoes or boiled turnips. It is my opinion that regularity in feeding is an exceedingly important point. Those, who throw down, at one time, double the amount of food the pigs can eat, and then let the proper time go by without feeding anything, find that their pork costs them double what it costs a careful and regular feeder who takes pleasure in watching his pigs eat."

James Howard, M. P., of Bedford, England, a very successful breeder of Large or Medium Yorkshires, writes: "Mr. Fisher, of Carhead, Yorkshire, has published a capital lecture he delivered on the management of pigs, which I will send you. I enclose also a letter from Mr. Sidney.

"The White Leicesters have disappeared. They had little or no hair. The large Yorks (not the Mammoth) are the most profitable as they grow so fast, and are turned into money quickly.

"There are no such animals as pure Suffolks. They are the Fisher Hobbs Essex variety.

"If you want sows to breed well, do not keep them too fat, nor yet in a weakly condition. Let them have a field to run about in. We used to fat a great many 'porkers' with pulped roots, straw, chaff, and Indian corn, but we have now such a large demand for breeding pigs, that we have none left for fatting. With respect to feeding, the food should be given warm, not hot."

From this last remark I conclude that the pulped roots, chaff, and Indian meal were cooked.

The following is an extract from Mr. Sidney's letter to Mr. Howard:

"I do not think our pigs have improved during the last ten years [since Mr. Sidney's book was written]. On the contrary, our Shows are likely to cultivate fat at the expense of constitution. I think Mr. Harris mistakes my advice. [I thought he condemned the use of an Essex or Berkshire boar with a *white sow*.] A cross of black and white answers well for feeding, as most first crosses do. I observe that black pigs have made their way in Yorkshire."

The following extracts are from Mr. John Fisher's lecture on the breeding and management of pigs, alluded to by Mr. Howard. Mr. Fisher is the manager of Carhead Farm, near Crosshills, Yorkshire, and an experienced and successful breeder. We should remark that Carhead is a grass farm, and all the food for the pigs is purchased. Mr. F. says:

"I am a decided advocate for early breeding and early feeding, and consider October or November the best time for putting sows to the boar for the general crop. They will then bring their litter in March, and get them weaned, and take the boar again early in May, so that their second litter may get strong enough to stand the winter; and if the young sows, bred in March, have been liberally fed, and allowed plenty of exercise during the summer, they will be quite ready to take the boar in November, and bring their first litter at twelve months old. And we consider this the best way either to commence or increase a stock of breeding pigs, and should not endorse the claim to early maturity in any breed of pigs, if they were not unfit to rear a litter of young at twelve months old. If young sows are allowed to run until they are twelve or fifteen months old before they are put to breeding, they are very apt to miss their way altogether; and we find that the most successful breeders are those which are put to, when young, and are kept regularly breeding; consequently we do not disappoint them, but allow them to bring two litters a year. After their first litter we keep them sparingly, except when suckling. When they have weaned their spring litter, and have taken the boar again, they are turned into a grass field, in which there is a large shed, with rails across the doorway to prevent cattle entering. In this shed they sleep at night, or retire in rainy weather. If the grass is not very plentiful, we give about a pint each of Indian corn per day, scattering it on the grass, and they can drink water from a stone trough which is fed by a spring, and placed near the ground that they may reach it conveniently. But they mostly gain so much flesh from being well fed while suckling, that they require little more than grass; and some which have had nothing else, have done as well as we could wish them.

"When the sows are brought into the breeding house, they are at once put on the same kind of food as will be continued to them while they are suckling. They are turned out for a few minutes twice a day, before feeding, which keeps the bowels in proper order, and the house dry and sweet, for it is very important that the bowels are open at this time, for, if constipated, the milk will not come freely, and the young seldom do well; besides which, it interferes with the free passage of the urine, causing great uneasiness, and, if not removed, it would lead to serious consequences, for which purpose we give frequent injections of warm water, and walk the patient carefully out, for a few minutes at a time, until we see that the obstruction has passed. Sometimes we mix a little common soap in the warm water, and have never experienced much difficulty when these means have been used.

"We give a moderate bed of short straw three or four days before they are expected to farrow, that it may become soft by the time they are due, which, as the time approaches, they will collect on a heap, and place themselves upon it in such a manner that by raising the body it assists them in their efforts during parturition, and this, as well as most other matters at this time, we leave entirely to themselves, believing that they can mostly manage their own business best without our interference. And except with very fat sows, or during very cold weather, we do not remain with them while farrowing, but give an occasional eye to them to see that there is no unusual delay. If the presentations are proper, they will often pass three or four in as many minutes, but when the hind feet are presented foremost, they get on slowly, and sometimes half the litter will come in this way, but assistance in such cases will mostly do more harm than good, for in drawing the birth by the hind legs, the viscera is forced into the chest, and the life is thereby endangered to no purpose, for if ever they get so far on their way as to be within the reach of ordinary aid, they will be passed safer without it.

"The pigs usually begin to eat along with the mother when about three weeks old, but may be learnt much younger if a little warm milk be given to them two or three times a day, while the sow is removed from them for a few minutes. About the time they begin to eat, they frequently suffer from diarrhœa, which, if it continue for any length of time, will weaken them very much. The disorder will sometimes be caused by allowing the mother to eat grass or other green food when turned out, or even by a change from one kind of meat to another, for which reason we avoid as far as we can any change of food during the time they are suckling, and continue the same to the young after they are weaned. And as it is very difficult as well as dangerous to administer medicine to them by force, we do not attempt to relieve them by that means, neither can they be induced to take it if mixed with their food, for they will not eat at such times, but depend entirely on the teat, for which reasons we diet the mother carefully, and allow as much small coal as she will eat, throwing a shovelful upon the bed, that the young ones may eat a little if they like; we also strew the floor with sawdust to prevent bad smells, keeping them warm, and giving as much fresh air as possible. If the purging continues, we change them to a fresh sty, taking care that it is dry and warm, and well aired. If young pigs can be allowed a run out with the mother for half an hour in the morning and evening, they will grow all the faster for it; but the middle of the day, when the sun is hot, should be avoided, for if their backs get much scorched it will retard their growth for a while.

"All such as are not required for breeding purposes, should be castrated at from four to five weeks old, that they may recover before they are weaned. There are two ways of doing most things, and the best way is generally the easier, and always to be preferred, and in catching young pigs for castration, or any other purpose, great care should be used, as they are easily lamed, and having covered the window and closed the door to exclude the light, the operator should allow them to settle quietly in a corner, and taking the right hind leg with his right hand, then with his left hand he should lay firmly hold of the *same leg*, above the hough joint, and quickly passing his right hand forward, and under the chest, lay firmly hold of the *left fore leg*, and raise the pig with his right hand, using as little force as possible on the hind legs, and never hold them up by the heels, as the intestines are liable to get twisted if held in that position.

"We usually wean at from seven to ten weeks old, and separate the boars from the sows soon after. We seldom keep more than five or six together in the same sty, and as they grow larger, we reduce the number, in proportion to the size of the sty.

"The feeder commences in the morning about seven o'clock, beginning at one end, and regulating the food according to circumstances, and as he goes on, he rouses every pig up, and sees that all come to take their breakfast; should any refuse he reports the case; and having finished feeding he takes his barrow, fork, shovel, and besom, and proceeds in the same order to clean the sties; for, on being roused up, after laying still all night, they empty themselves while eating, and this becomes habitual and keeps their beds clean and dry, which is a matter of great importance to us, as we have all our straw to buy at a dear rate, and have to economise it accordingly, for which reason most of our sties are provided with wooden sleeping benches similar to that given in the description of the breeding house. So proceeding to No. 1, he turns the occupants out, shakes up the bed, sweeps all clean, and taking up with the shovel what had to be removed, he places it in the barrow, returns them to their sty again, and passing on to No. 2, treating them in the same way, and so on to the end. By this means the sties are kept clean during the greater part of the day; while out, they have free access to a heap of small coal, which is kept in a corner of the yard entirely for their use, of which they seldom fail to avail themselves, whenever they have an opportunity; there is also a trough with water, of which they sometimes drink a little.

"To enable pigs to thrive properly, they must be kept in a state of robust health, for which purpose, proper shelter and a certain amount of exercise, is quite as necessary as good feeding, and all dark, damp, crampy sties should be avoided. There is no place in which young growing stores do better than a good straw-yard during the winter months.

"Pigs will occasionally catch cold, especially when in low condition; but, if taken in time, and placed in a warm sty by themselves, with a little extra nursing, such as warm milk and water, with a little bran or pollard, not forgetting the warm water injections if the bowels get out of order, they will mostly be right again in a few days. If the case be a bad one, and accompanied by much fever, and the patient will lie still, we cover up with a wet rag, leaving only the nose out, pouring cold water on to saturate it thoroughly, and then cover up with two or three sacks to keep the steam in, and have found this bath give very great relief. Pigs have a very great objection to any kind of restraint, as well as a strong dislike to physic, and if held for the purpose of administering it, they struggle and scream so much, that they do themselves more harm by it, than the medicine is likely to do them good; besides, if not done in a careful manner, there is great danger in forcing any liquid into their mouths, for if introduced while they are screaming, they are almost certain to be choked by it, so that the operator must wait until they have done screaming, which will mostly be when they are out of breath and cannot go on any longer, for which reason we have not used medicine for several years past. They have also a very decided objection to strangers being admitted into their society, even if one of their fellows leave them for a few days, on their return they are beset and worried in a most unfriendly manner; and if the intruder cannot find means of retreat, they will often get cut and gored a good deal; where the teeth penetrate beyond the skin, swellings will arise, which if they become very large, they may be carefully opened with a lance, or sharp pointed knife, on the lower side, directing the point upwards, that the matter may escape, when they will soon heal without further trouble.

"Fat heavy pigs are easily lamed in the hind-quarters or hind legs, and should be very carefully driven over slippery or uneven ground. When so lamed, the butcher is the best remedy and the sooner the better, as they lose flesh fast, when they come to lie and cannot rise easily. They are also subject to rheumatic attacks, especially in the hind legs, which may easily be mistaken for accidental lameness; sometimes they will suddenly become lame in one leg, and then the lameness will as suddenly change to the other, or perhaps leave them altogether. I consulted Professor Simonds, of the Royal Veterinary College, on this disease, and he recommended a strong stimulating liniment, or liquid blister to be applied to the hough joint, and well rubbed in, and I have used it with very beneficial results; also, if confined for any length of time where the wet litter is allowed to accumulate under them, their hoofs grow to a great length, and the feet become unsound and full of clefts, when the hoofs should be shortened, for which purpose we use a pair of strong, wire-clipping pinchers, taking care not to injure the sensitive part of the foot, and trim with a shepherd's knife; and for diseased feet we have found nothing so good as a bran poultice, with two or three spoonfuls of fresh brewer's yeast mixed with it, and put in a strong bag or boot, into which the foot is introduced, and secured with a string when the animal is laid down. It may be kept wet by pouring water on it two or three times a day, and changed daily."

T. L. Harison, Esq., Morley, St. Lawrence Co., N. Y., writes:

"I do not think I can give you any ideas of value as to the breeds and breeding of pigs, for my experience has been with Suffolks only, and the breeding them has been with me a matter of great simplicity, and in which I have found no difficulties to contend with. I have found the Suffolks hardy, prolific, good nurses, and good feeders. Those who have had barren sows have, I think, allowed them to get too fat before breeding. This is the only risk that I know of, and it is to be guarded against. My plan was to keep over such young sows as I selected for breeders generally from fall litters, but seldom from spring litters. These were usually kept in a yard or in a small grass field, so that they were on the ground and had plenty of exercise, and when served about December 1st would be from 14 to 16 months old and in fair (*extra*, perhaps) store condition. After they were with pig, they would of course during the winter get fat, but in my breeding that never did any harm. My only trouble was in the loss of young pigs, in consequence of the milk of their mothers being too rich. This makes it necessary to be careful how you feed the sows while suckling, and I found that bran with the refuse of the house made a better food than grain at such times.

"I do not know about plans of pig pens. I have never seen any that I thought had much merit. In fact, I would never use pens, except for fattening hogs, for the boars in use, and occasionally for breeding sows before farrowing; but, except in the first case, there should be small yards attached. The best place for pigs is a yard with a well-made shed attached, the shed having doors that can be closed in very severe weather."

Hon. John M. Milliken, of Maplewood, Hamilton, Ohio, in addition to the facts already quoted in regard to the Butler County or Magie pigs, writes as follows: "I wish to add the following statement furnished me by one of our breeders, whose truthfulness is unquestioned. He bred a sow which came on the 10th of June, 1866. On the 18th of April, 1867, she had 11 pigs, which weighed, gross, in October following, 2,735 lbs.

"He fattened the sow the winter following, and her net weight was 535 lbs. The sow pigs he left for breeders, and sold 5 barrows, aged 8 months and 20 days, which averaged 282 lbs. net. The history of this sow and her 11 pigs proves that they possessed early fattening properties, large size, and fecundity,—three very desirable qualities."

INDEX.

Allen's, B. A., importation of Berkshire...... 99
Appendix........................237
Bakewell's, Robert, breed of pigs.. 18
Barley, Composition of............133
" Value in feeding..........131
Bean meal, Value in feeding.......126
Beans as food for pigs............233
" Composition of............133
Bedfordshire breed................. 93
Berkshire breed............50, 74
" A. B. Allen's importation. 99
" and Tamworth cross..... 83
" Crosses of................ 83
" Improved................ 76
" in the U. S............. 98
Black and Red pigs................ 86
Black breeds...................... 73
Blister for pigs...................234
Boar, Selection of................192
" Treatment of thorough-bred.206
Boussingault's experiments in feeding............118
Bran, Composition of.............133
" for pigs.....................239
" Value in feeding............126
Breeding and management of pigs 192, 239
Breeding, Objects in............... 15
" Principles of........... 16
Breeding-pens.....................200
Breeds of pigs..................... 14
Breeds. Definition of large, small, and medium................ 25
Breeds, Large vs. small.............. 22
" Leading English..........232
" Leading in the U. S......232
" Modern English........... 56
" in the U. S................ 98
" Sidney on large and small.. 82
Buckinghamshire breed............ 97
Bushey breed...................... 97
Butler Co., Ohio, pigs.............113
Cattle, Gain from feeding.......... 18
" Weight, Live and dead......192
Castration, Time for..............233
Chase's, A. L., Essex pigs........... 84

"Cheshire" breed..........57, 94, 111
Chester Co. White breed.........105
" " " " Hog-breeders' Manual on.110
" " " " Paschal Morris on......................106
Chinese breed..................... 51
Codfish, Composition of..........133
" Value in feeding...........133
Coleshill breed.................... 96
Colds in young pigs...............217
Compost, Pig making..............142
Constipation......................210
Cooking food for pigs.............232
Craonnaire Boar................... 45
Crosses of thorough-breds......... 85
Culley on Cheshire pigs........... 57
Cumberland small breed........... 68
Davis', Hewitt, experience in pig feeding..................... 90
Desirable qualities in a pig........ 20
Devons.........................83, 87
Diarrhœa in young pigs...........217
Diseases of pigs..................234
Disinfectants.....................234
Dorsets........................33, 88
Dry earth for pigs................234
"Emperor."........................ 53
English breeds, Improvement of... 47
English experiences in pig feeding.181
Essex, Fisher Hobbs' improvements in..................53, 82
Essex, Improved................... 80
" " Crosses of......... 82
" " History of......... 52
" Imported..................100
" Lord Western's............. 82
" Old......................... 52
Experiments in pig feeding....118, 122
Fancy breeds...................... 95
Fattening pigs near large cities....178
Feeding, A dairy farmer on........187
" A Yorkshire breeder on..188
" A Yorkshire farmer on...186
" Dr. M. Miles' experiments 119
" English experience in....181
" Experiments in..........118

Feeding Grain................225
" Lawes and Gilbert's experiments in............122
" Mineral substances necessary in................130
" Mr. Baldwin on..........182
" Use of sugar in............135
Feet, Unsound....................244
Fisher Hobbs' improvements in Essex........................53
Fisher, John, Lecture on breeding and management.............239
Food, Cooking...................222
Form of a fat pig................296
" of a good pig..............17
French pigs......................45
"General."......................68
German pigs....................45
Gestation, Period of............283
"Gloucester."....................67
Good pigs need good care........37
Grade pigs, Value of...........100
Greyhound hog..................44
Hampshire pig.............49, 91
Herefordshire breed..........50, 94
Howard, James, M. P., on pigs....239
Improvement of the English breeds of pigs....................47
In-and-in breeding..............35
Indian meal, Composition of....133
" " Value in feeding.....126
Intestines, Proportion to weight of body........................11
Itch...........................234
Jealousy an aid to fattening....230
Jefferson County breed.........111
Large vs. small breeds and crosses. 22
Lawes and Gilbert's experiments in pig feeding..............122
Lecture by John Fisher..........239
Lentil meal, Value in feeding...126
Lentils, Composition of.........133
"Liberator."....................67
Lice, To destroy................234
Lincolnshire breed..........57, 92
Liquid manure..................148
Live and dead weight of pigs....191
Lord Western's Essex............52
Mackelcan, J., on management....237
Magie (Ohio) breed..........118, 245
Management of pigs..........175, 237
" of thorough-bred pigs.203
Mange..........................234

Mangles', George, experience in feeding......................95
Mangles', George, piggery.......163
Manure, Table of value of......139
Manure, The pig as a manufacturer of........................141
" Value of liquid..........148
" Value to each 100 lbs. of pork..................141
" Value of pig..............137
Measles........................235
Michigan Agricultural College, Piggery, etc...................147
Middlesex breed................95
Miles' Dr. M. experiments in feeding118
Milliken, Hon. J. M., on Magie pigs....................118, 245
Mineral food for pigs..........130
"Miss Emily."..................70
Modern breeds of English pigs....56
Morris', Paschal, piggery......154
Neapolitan breed...............52
Norfolk breed.................93
Nottinghamshire breed..........95
Ogden Farm piggery............160
Oil-cake for pigs..............233
Old Irish pig..................44
Old Yorkshire breed............57
Original Old English pig.......43
Origin and improvement of our domestic pigs...................41
Oxfordshire, Improved..........85
Ox, Stomach of.................9
Peas for pigs..................233
" Raising, for pigs 177
Pen-breeding...................200
Petroleum on pigs..............234
Pig, Desirable Qualities in.....20
" Form of a good..............17
" Quietness in................21
" Stomach of..................9
Pig feeding....................11
" " Hewitt Davis on........30
" " Why they gain more rapidly than oxen or sheep....12
Pigs on dairy farms............175
" on grain farms..............176
" Origin and improvement of..41
" Peas for....................177
" Profit of raising thorough-bred.......................220
" require gentle treatment....42
" Breeding and rearing........192

Pigs, Breeds of.................... 14
" Cooking food for.............222
" Fattening near large cities...178
" Lame.. 244
" Management of.............175
" Management of thorough-bred 203
" Young, Care of.........196, 242
" " Catching........ 242
" " Colds in..............217
" " Diarrhœa in.....217, 243
" " Feeding...... 212
" " Management of.......217
" " Taming 219
" " Time to wean...197, 242
" " Treatment of chilled.212
Piggery, The author's..............148
" George Mangles'.... 163
" Michigan Agricultural College...................147
" Mr. Roseburgh's..........158
" Ogden Farm..............160
" Paschal Morris'..........154
" Tattenhall (Eng.).........166
Piggeries and pig pens.......... .. 144
Pig pens.........................144
" Location of.................144
Pig troughs......................169
" " Cast-iron 174
" " Convenient......... 172
" " hewn out of a log.......170
" " Plank............. 171
" " Swinging door..........173
Pork, Food required to produce 100 lbs...................... 11
Prince Albert's pigs 96
Profit of raising thorough-bred pigs 222
Protrusion of rectum..............236
Pulping roots.... 228
Rectum, Protrusion of..............236
Rheumatism....... ....236, 244
Roseburgh's, Mr., piggery...... ..158
Sheep, Lawes' experiments in feeding..... 22
" Stomach of.................. 9
" Live and dead weight of.....192
Shropshire breed..................... 94
Sidney on large and small breeds.. 32
Soaking grain for pigs........ 228
Sow at farrowing time.........194, 241
" Breeding, Management of.....241
" Feeding a suckling............214

Sow lying on pigs................211
" Selection of....................193
" taking the boar...............233
" Treatment of thorough-bred..209
Spring pigs, Rearing and management of........................200
Stickney's, Isaac and Josiah, importation of Suffolk.... 100
Stomach, Importance of a good.... 20
" of ox, Weight of......... 9
" of pig, Weight of......... 9
" Proportion to weight of body............... 11
" of sheep, Weight of...... 9
Stone, F. W., on pigs......... 238
Suffolk and other white breeds..... 72
Suffolk breed..................... 92
" grades....................103
" introduced into Boston... 100
Sugar as food for pigs............135
Summary.........................232
Sus Indica........................ 41
Sus scrofa........................41
Swellings, Treatment of...........243
Swill barrels... ... .169
Swill barrel, Portable.............170
Swill tub.........................170
Tamworth breed.......... .....86, 87
Tamworth and Berkshire cross.... 88
Tapeworm......... 235
Tattenhall piggery....... 166
Thorough-bred pig, Value of..... . 85
Value of a thorough-bred pig... .. 85
" of pig manure............ ...137
Warwickshire breed............... 86
Weight of pigs, Live and dead... .191
" of different parts of a pig..191
Weaning young pigs...............197
Welsh pigs........ 94
White Leicesters................59, 72
Wild boar...................... 41
Wild hogs................... 41
Windsor breed................... 96
"Windsor Castle"................. 99
Woburn breed....... ....33, 94
Yelt. 66
York-Cumberland breed........... 65
Yorkshire grades..................103
" introduced into the U. S.100
" large breed............. 59
" middle or medium breed. 60
" small breed............. 63

LIST OF ILLUSTRATIONS.

Diagram of, Testing the Form of a Good Pig.. ....Page 18
Wild Boars.... 42
Wild Boar... 43
Original Old English Pig.. 43
Old Irish Pig....... 44
French Prize Boar—Craonnaire White.. 46
Chinese Sow—Imported... 48
Berkshire Pig—(*Loudon*).. 49
Hampshire Pig   " 49
Herefordshire Pig   " 50
Suffolk Pig   " 51
Berkshire Sow... 51
Yorkshire Large Breed, "Sir Roger de Coverly."............................ 58
Yorkshire Large Breed, "Parian Duchess."................................. 60
Yorkshire Large Breed, "Golden Days."..................................... 62
Cumberland-York Boar—Small Breed.. 64
York-Cumberland Pig—Small Breed... 66
Yorkshire Middle Breed—"Miss Emily."...................................... 68
White Leicester Boar and Sow—Small Breed...... 76
Berkshire, Improved—Smithfield Club Fat Prize Sow......................... 75
Berkshire, Improved, Middle Breed, Boar................................... 77
Essex, Improved—"Emperor.".. 79
Essex, Lord Western's... 87
Essex Boar, L. A. Chase's... 83
Essex Sows   "   " 83, 84
Chester County White Pig..105
Jefferson County Pig..111
Piggery, Michigan Agricultural College....................................147
Piggery, Ground Plan of Michigan Agricultural College.....................148
Piggery, Plan of the Author's...151
Piggery, Paschal Morris'..154
Piggery, Ground Plan of Paschal Morris'...................................155
Piggery, Mr. Roseburgh's..157
Piggery, Ground Plan of Mr. Roseburgh's...................................159
Piggery, Partition in   "   " 160
Piggery, Ground Plan of Ogden Farm..161
Piggery, Cross Section of Ogden Farm...................... 162

Piggery, Elevation of Ogden Farm...162
Shed for Fattening Pigs, Mr. Mangles'..163
Shed, Ground Plan of   "   " 163
Shed, Isometrical view of "   " 165
Piggery, Covered Food House at Tattenhall...............................166
Piggery, Ground Plan of Covered Food House at Tattenhall...,..........167
Portable Swill Barrel...170
Pig Trough—Hewn Out of a Log 170
Pig Trough, Plank..171
Pig Trough, Convenient..172
Pig Trough, Swinging Door...173
Pig Trough, Swing Door to..174
Pig Trough, Cast-iron..174
Brood Sow—Property of the Duke of Buccleuch............................180
Jealousy as an Aid to Fattening..281

AMERICAN CATTLE:
Their History, Breeding, and Management.
By LEWIS F. ALLEN,
Late President New-York State Agricultural Society, Editor "American Short-Horn Herd Book," Author "Rural Architecture," etc., etc.

Notices by the Press.

WE consider this the most valuable work that has recently been issued from the American press. It embraces all branches of the important subject, and fills a vacancy in our agricultural literature for which work the author, by his many years' experience and observation, was eminently fitted. . . . It ought to be in the hands of every owner of cattle, and the country, as well as individuals, would soon be much richer for its teachings.—*Journal of Agriculture*, (*St. Louis.*)

The large experience of the author in improving the character of American herds adds to the weight of his observations, and has enabled him to produce a work which will at once make good its claims as a standard authority on the subject. An excellent feature of this volume is its orderly, methodical arrangement, condensing a great variety of information into a comparatively small compass, and enabling the reader to find the point on which he is seeking light, without wasting his time in turning over the leaves.—*N. Y. Tribune.*

This will rank among the standard works of the country, and will be considered indispensable by every breeder of live-stock.—*Practical Farmer*, (*Phila.*)

We think it is the most complete work upon neat stock that we have seen, embodying as it does a vast amount of research and careful study and observation.—*Wisconsin Farmer.*

His history of cattle in general, and of the individual breeds in particular which occupies the first one hundred and eighty pages of the volume, is written with much of the grace and charm of an Allison or a Macaulay. His description of the leading breeds is illustrated by cuts of a bull, a cow, and a fat ox, of each race. The next one hundred pages are devoted to the subject of Breeding. This is followed by chapters on Beef Cattle, Working Oxen, Milch Cows, Cattle Food, Diseases, etc. The arrangement, illustrations, analytical index, etc., of the work are in the best style of modern book-making.—*New-England Farmer.*

The work is one that has been long needed, as it takes the place of the foreign books of like nature to which our farmers have been obliged to refer, and furnishes in a compact and well-arranged volume all they desire upon this important subject.—*Maine Farmer.*

Whatever works the stock-farmer may already have, he can not afford to do without this.—*Ohio Farmer.*

It is one of the best treatises within our knowledge, and contains information sound and sensible on every page.—*The People*, (*Concord, N. H.*)

The object of the work, as stated by the author in his preface, "is not only to give a historical acccount of the Bovine race, to suggest to our farmers and cattle-breeders the best methods of their production and management, but to exalt and ennoble its pursuit to the dignity to which it is entitled in the various departments of American agriculture." From the little examination we have been able to give it, we can not recommend it too highly.—*Canada Farmer.*

Considering that there are some ten million milch cows in the United States, and nearly a thousand million of dollars invested in cattle, the magnitude of this interest demands that the best skilled talent be devoted to the improvement of the various breeds and the investigation of the best method of so caring for the animals as to gain the greatest profit from them. This volume will give the farmer just the instruction which he wants.—*N. Y. Independent.*

Price, post-paid, $2.50.

ORANGE JUDD & CO.,
245 Broadway, New-York.

NEW AMERICAN FARM BOOK.

ORIGINALLY BY

R. L. ALLEN,

AUTHOR OF "DISEASES OF DOMESTIC ANIMALS," AND FORMERLY EDITOR OF THE "AMERICAN AGRICULTURIST."

REVISED AND ENLARGED BY

LEWIS F. ALLEN,

AUTHOR OF "AMERICAN CATTLE," EDITOR OF THE "AMERICAN SHORT-HORN HERD BOOK," ETC.

CONTENTS:

INTRODUCTION.—Tillage Husbandry—Grazing—Feeding—Breeding—Planting, etc.

CHAPTER I.—Soils—Classification—Description—Management—Properties.

CHAPTER II.—Inorganic Manures—Mineral—Stone—Earth—Phosphatic.

CHAPTER III.—Organic Manures—Their Composition—Animal—Vegetable.

CHAPTER IV.—Irrigation and Draining.

CHAPTER V.—Mechanical Divisions of Soils—Spading—Plowing—Implements.

CHAPTER VI.—The Grasses—Clovers—Meadows—Pastures—Comparative Values of Grasses—Implements for their Cultivation.

CHAPTER VII.—Grain, and its Cultivation—Varieties—Growth—Harvesting.

CHAPTER VIII.—Leguminous Plants—The Pea—Bean—English Field Bean—Tare or Vetch—Cultivation—Harvesting.

CHAPTER IX.—Roots and Esculents—Varieties—Growth—Cultivation—Securing the Crops—Uses—Nutritive Equivalents of Different Kinds of Forage.

CHAPTER X.—Fruits—Apples—Cider—Vinegar—Pears—Quinces—Plums—Peaches—Apricots—Nectarines—Smaller Fruits—Planting—Cultivation—Gathering—Preserving.

CHAPTER XI.—Miscellaneous Objects of Cultivation, aside from the Ordinary Farm Crops—Broom-corn—Flax—Cotton—Hemp—Sugar Cane—Sorghum—Maple Sugar—Tobacco—Indigo—Madder—Wood—Sumach—Teasel—Mustard—Hops—Castor Bean.

CHAPTER XII.—Aids and Objects of Agriculture—Rotation of Crops, and their Effects—Weeds—Restoration of Worn-out Soils—Fertilizing Barren Lands—Utility of Birds—Fences—Hedges—Farm Roads—Shade Trees—Wood Lands—Time of Cutting Timber—Tools—Agricultural Education of the Farmer.

CHAPTER XIII.—Farm Buildings—House—Barn—Sheds—Cisterns—Various other Outbuildings—Steaming Apparatus.

CHAPTER XIV.—Domestic Animals—Breeding—Anatomy—Respiration—Consumption of Food.

CHAPTER XV.—Neat or Horned Cattle—Devons—Herefords—Ayrshires—Galloways—Short-horns—Alderneys or Jerseys—Dutch or Holstein—Management from Birth to Milking, Labor, or Slaughter.

CHAPTER XVI.—The Dairy—Milk—Butter—Cheese—Different Kinds—Manner of Working.

CHAPTER XVII.—Sheep—Merino—Saxon—South Down—The Long-wooled Breeds—Cotswold—Lincoln—Breeding—Management—Shepherd Dogs.

CHAPTER XVIII.—The Horse—Description of Different Breeds—Their Various Uses—Breeding—Management.

CHAPTER XIX.—The Ass—Mule—Comparative Labor of Working Animals.

CHAPTER XX.—Swine—Different Breeds—Breeding—Rearing—Fattening—Curing Pork and Hams.

CHAPTER XXI.—Poultry—Hens, or Barn-door Fowls—Turkey—Peacock—Guinea Hen—Goose—Duck—Honey Bees.

CHAPTER XXII.—Diseases of Animals—What Authority Shall We Adopt?—Sheep—Swine—Treatment and Breeding of Horses.

CHAPTER XXIII.—Conclusion—General Remarks—The Farmer who Lives by his Occupation—The Amateur Farmer—Sundry Useful Tables.

SENT POST-PAID, PRICE $2.50.

ORANGE JUDD & CO.,

245 Broadway, New-York.

www.ingramcontent.com/pod-product-compliance
Lightning Source LLC
Chambersburg PA
CBHW080523240526
45472CB00021BA/1752